Advances in Biology

nova
science publishers
www.novapublishers.com

Advances in Biology

Advances in Biology. Volume 5
Charles D. Grant (Editor)
2024. ISBN: 979-8-89113-441-6 (Hardcover)
2024. ISBN: 979-8-89113-475-1 (eBook)

Advances in Biology. Volume 4
Charles D. Grant (Editor)
2024. ISBN: 979-8-88697-999-2 (Hardcover)
2024. ISBN: 979-8-89113-074-6 (eBook)

Advances in Biology. Volume 3
Charles D. Grant (Editor)
2024. ISBN: 979-8-88697-779-0 (Hardcover)
2024. ISBN: 979-8-88697-846-9 (eBook)

Advances in Biology. Volume 2
Charles D. Grant (Editor)
2024. ISBN: 979-8-88697-422-5 (Hardcover)
2024. ISBN: 979-8-88697-435-5 (eBook)

Advances in Biology. Volume 1
Charles D. Grant (Editor)
2024. ISBN: 979-8-88697-201-6 (Hardcover)
2024. ISBN: 979-8-88697-218-4 (eBook)

More information about this series can be found at
https://novapublishers.com/product-category/series/advances-in-biology/

Charles D. Grant
Editor

Advances in Biology

Volume 6

nova
science publishers
www.novapublishers.com

NOTICE TO THE READER

Library of Congress Cataloging-in-Publication Data

ISBN: 979-8-89113-712-7 (Hardcover)
ISBN: 979-8-89113-790-5 (eBook)
ISSN: 2833-1753

Published by Nova Science Publishers, Inc. † New York

Contents

Preface	 vii
Chapter 1	**Expansion of R *Radiobacter* Strain R58. Ti-plasmid – Bio Time Bomb** .... 1 *Kristina Zubow, Anatolij Zubow and Viktor A. Zubow*
Chapter 2	**Effects of Methyl Tertiary-Butyl Ether (MTBE) on Testis Tissue and Sex Ratio of Offspring in Adult Rats** .... 23 *Leila Khalili and Abbas Khalili*
Chapter 3	**Lactic Acid Bacteria in the Gastrointestinal Tract: Anti-*Helicobacter pylori*** .... 37 *Sujitra Techo, Engkarat Kingkaew and Somboon Tanasupawat*
Chapter 4	**Phytoremediation Potential of *Azolla* Species: A Comprehensive Review of Physiological, Biochemical, and Molecular Responses to Heavy Metal Contamination in Freshwater Environments** .... 65 *Seran Kirkinci, Yusuf Can Gercek and Taylan Kosesakal*
Chapter 5	**Algal Lipid Metabolism: Recent Advances in Growth and Adaptation** .... 103 *Yigit Sabri Unlu, Hande Morgil, Yusuf Can Gercek and Taylan Kosesakal*

Chapter 6 **Uncovering the Botanical Alarms: Electrical Signaling in Response to Insect Herbivory - A Review** 133
Gabriela Niemeyer Reissig, Thiago Francisco de Carvalho Oliveira, Douglas Antônio Posso, Helena Chaves Tasca, Luis Felipe Basso, Lyana Pintos Ramos, João Gabriel Moreira de Souza, André Geremia Parise and Dori Edson Nava

Chapter 7 **Comparison of the Basic Chromosome Topologies of Onion (*Allium cepa*) from Germany and China** 177
Kristina Zubow, Anatolij Zubow and Viktor A. Zubow[1]

Index 195

Preface

This book is a compilation of seven chapters on the advancements in biology. Chapter one assesses the physiological, biochemical, and molecular impacts of environmental conditions on microalgal lipids and lipid metabolisms. Chapter two evaluates the physiological, biochemical, and molecular impacts of HMs contamination on Azolla species, shedding light on the potential application of this plant for phytoremediation purposes. Chapter three aims to provide a comprehensive review of electrical signals associated with herbivory in plants, encompassing ecological perspectives, signal types, generation mechanisms, detection, analysis techniques, and future applications. Chapter four highlights the diversity of gut microbiota and lactic acid bacteria present in the GI tract and reviews important findings associated with alterations of gut microbiota in H. pylori-infected humans and mice, and clinical trials that use probiotics and bacteriocins as alternative or adjunctive therapies for the treatment of H. pylori infection. In chapter five the expansion of R radiobacter strain R58 in the eukaryotic biocenosis in northern Germany is studied using gravitational mass spectroscopy (GMS). The authors in chapter six study the effect of oral exposure to MTBE on histologic and histomorphometric changes of testis in adult rats and investigate the effect of MTBE on offspring sex ratio, sex hormones and antioxidant enzyme. In the final chapter, the basic topology of the molecular architecture of the chromosomes of the onion Allium cepa from China and northern Germany is studied using the gravitational mass spectroscopy (GMS) method.

Chapter 1

Expansion of R *Radiobacter* Strain R58. Ti-plasmid – Bio Time Bomb

Kristina Zubow[1,*]
Anatolij Zubow[2]
and Viktor A. Zubow[1]

[1]R&D, Zubow Consulting (NPO), Waren, Germany
[2]Dept. Telecommunication Networks Group, TU Berlin, Berlin, Germany

Abstract

The expansion of R *radiobacter* strain R58 in the eukaryotic biocenosis in northern Germany was studied using gravitational mass spectroscopy (GMS). Anhydrides of Ti-bacterial plasmid were found in biomatrices of almost the entire biocenosis. The source of spread is animals that use cereals as food, especially cattle. The use of manure as fertilizer on agricultural land leads to the entry of Ti-plasmid into the food chain of humans and domestic animals. Ti-plasmid was found in the cytoplasm of the liver, dermis of the skin, as well as in the micro biocenosis of human saliva. Using cereals as an example, a method of Ti-plasmid destruction was proposed. The presence of Ti-plasmid anhydrides in eukaryotic tissues indicates the absence of free water in them. Removal of water by drying or freezing at 233 K leads to the appearance of anhydride forms of Ti-plasmid. A conclusion is drawn about the potential danger of the expansion of the bacterium and its genome for humans.

Keywords: R *radiobacter*, expansion, cytoplasm, water, eukaryotes, Ti-plasmid, danger

---

* Corresponding Author's Email: vzubow@googlemail.com.

In: Advances in Biology. Volume 6
Editor: Charles D. Grant
ISBN: 979-8-89113-712-7

1. Introduction

Understanding the state of Ti-plasmid and water, its long-range order (LRO) [1], in the cytoplasm of eukaryotic cells is an important task for the correct interpretation of the results of biochemical studies. Here a problem arose with the indication of cluster-free water. Such an indicator can be Ti-plasmid anhydrides, which, as it turns out, are widespread in eukaryotic milieus [2]. On the other hand, the T-plasmid can integrate into a gene and cause cancer in plants and beyond, with a potential threat to the genomes of animals and humans. Let's try to test this hypothesis using gravitational mass spectroscopy (GMS).

The purpose of this work was to study the state of Ti-plasmid and its anhydrides, long-range order in the water of the cytoplasm of eukaryotic cells.

2. Material and Method

Eukaryotes (see below in the text), as well as samples of human dermal cells, micro biocenoses of saliva were chosen as the object of study. The micro biocenoses of cattle manure were studied. All samples are from the northeastern part of Germany. The state of water was studied using gravitational mass spectroscopy (GMS), [1]. The gravitational sensor was immersed in the sample and recorded gravitational noise (GN) from atomic nuclei clusters (ANC) in the mass range from 200 to 3.6 billion Daltons. GMS spectra were obtained using the first Zubow equation [1], the Zubow force constant (**Z**) was taken for water **Z** = 6.4E-12 N/m, for T-DNA, Ti-plasmid R *radiobacter* strain R58 (https://ru.wikipedia.org/wiki/Ti-plasmid), **Z**=6.39112655E-12 N/m (214,233 bp or 137,751,819 Da, 643 = 1 bp). Positive values of Δf are characteristic of loose forms of genomes; they reflect the energy part of genomes in the entire ensemble of studied ANC; negative values of $-\Delta f$ reflect the same thing, but for dense conformations of genomes. The average mass of the ANC in the ensemble, $M_{GMS} = \sum abs(\Delta f \cdot m)$, N is the number of signals from the ANC in the spectrum, D_c is the part of signals from dense, rich in potential energy ANC. Scanning time 10 s. Statistical analysis was carried out by studying ~2 thousand GMS spectra. The object of the study was the grains of the following cultivated seeds of cereal plants: wheat (Triticum, https://ru.wikipedia.org/wiki/Wheat), rye (Secále cereále, https://ru.wikipedia.org/wiki/Rye), oats (Avéna, https://ru.wikipedia.org/wiki/Oats), barley (Hórdeum, https://ru.wikipedia.org/wiki/Barley) cultivated in 2023 in the north of Germany (Upper Pomerania). This group also included wild varieties of rye and oats found on the territory of Earth, as well as samples of oak, vegetables, freshwater mollusks, reptiles, insects, microbiocenoses of cattle manure, saliva of the probant, dermis of the skin of the probant, and liver of the probant.

3. Results and Discussion

To search for Ti-plasmid signals in plants and calibrate the device, an oak tree with an obvious tumor lesion in the lower part of the trunk was selected, Figure 1.

As can be seen from the spectra of this Figure, in the area of the tree tumor, Ti plasmid signals are detected in a dense (**A**) and loose (**B**) conformation of the ring, like an oscillator. There are no plasmid signals in sample **B**, which indicates the recessive nature of the old tumor. The plant rejected the part of its tissue affected by the tumor and thereby isolating the healthy part of the tree from damage by the bacterium; here the immune system destroyed the genome of R radiobacter strain R58.

Figures 2-4 show fragments of the GMS spectra of cereal seeds in the region of oscillations of the genome of R *radiobacter* strain R58.

As can be seen from the spectra of Figure 2, in the region of expected oscillations of the DNA double helix ring with a mass of 137,751,854 Da, there is no signal (**I**). The wild specimen of the plant was very dry by this time. Therefore, searches for the genome ring signal were carried out in a different range of gravitational noise. It turned out that the genome ring loses two types of water (**II**), giving two signals for the oscillating masses: 129,574,300 Da (A, loss of 2.12 molecules of water per repeating base (bp) in DNA) and 130,378,654 Da (B, loss of 1.91 molecules of water per one repeating base in DNA).

Thus, the genome of R *radiobacter* strain R58 is represented by a ring of double helix DNA with two types of adsorption water, upon desorption of which two new types of rings of double helixes of DNA appear. Let's call them genome anhydrides A and B.

In cultivated oat (freshly cut stem), the genome ring (137,751,854 Da) oscillates like a pendulum in a loose conformation (**I**). Dehydration of oat grains under mild conditions (293 K) already reveals 3 types of anhydrides: A, B and C (**II**), with A in a loose conformation, and B and C in dense forms (Figure 3).

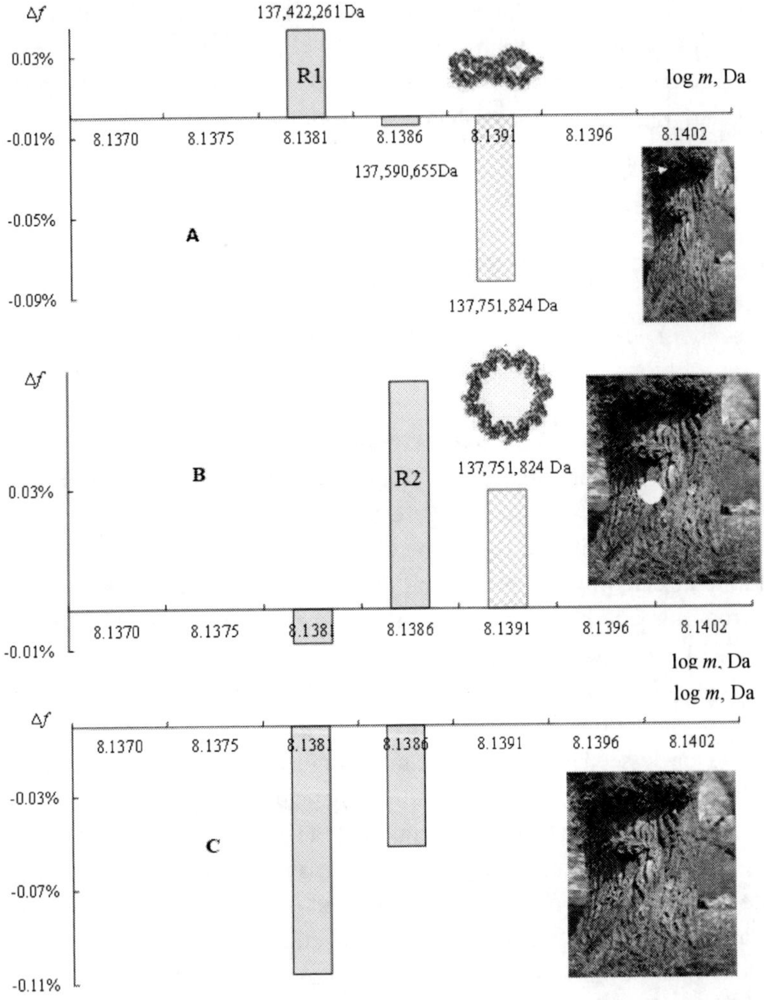

Figure 1. Fragment of the GMS spectrum of the oak biomatrix in the region of Ti-plasmid oscillations (signal of a pendulum with a mass of 137,751,824 Da). The sampling site is indicated by an arrow. **A** is the sampling site (bark) in the area of the tumor, **B** is the sampling site outside the tumor area (lower bark), and **C** is the sampling site in the area of healthy tree tissue (at a depth of 2 cm). Oak from the region of cultivation of cereals. Signals from the masses 137,422,261 and 137,590,655 Da are the reference signals of the ANC biomatrix. R1 and R2 are reference signals for calibrating the device (not indicated further in the text).

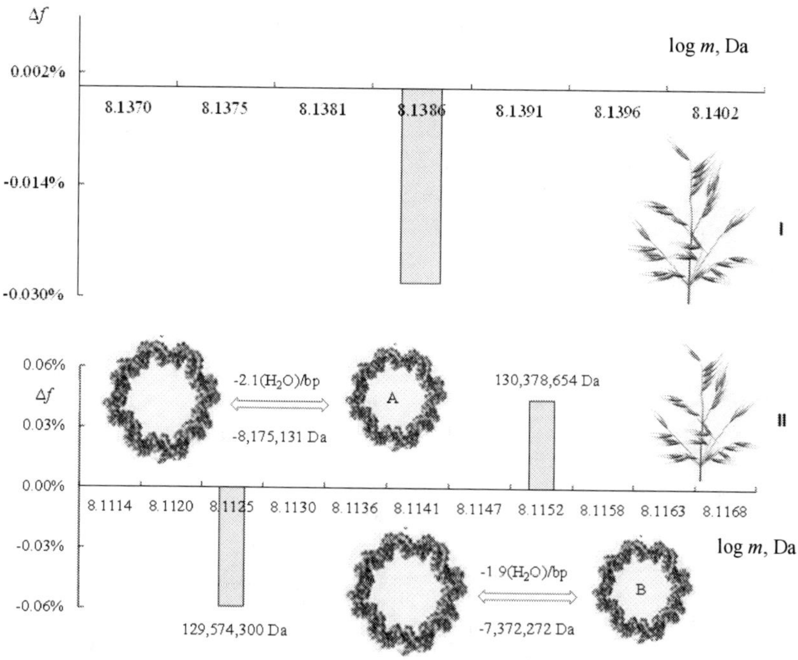

Figure 2. Fragments of the GMS spectra of the cytoskeleton of wild oats (grain, https://ru.wikipedia.org/wiki/Oats) in the expected region of oscillations of the hydrated Ti plasmid (**I**) and in the region of oscillations of the dehydrated plasmid (**II**). 07/05/2023.

In the fresh grain of wheat and rye (Figure 4), a genome signal with a mass of 137,751,854 Da is detected, which is presented in loose conformations, and in the wild rye variety in a dense form. At the same time, no anhydride signals were recorded in cultivated varieties of wheat, and in wild rye all three types A, B and C, as in the case of oats, Figure 3, **II**.

The grain drying process can also occur as a result of water freezing, for example, at low temperatures.

Figure 5 shows fragments of the GMS spectra of ANC oscillations in the cytoskeletons of barley grains exposed to low temperatures. It can be seen that in the original sample the signal of Ti-plasmid oscillations is very weak (137,751,824 Da), in it the equilibrium is shifted towards the dominance of dense conformations of rings of DNA double helices (**I**). Keeping the grain for 24 hours at 233 K leads to the disappearance of signal (**II**) as a result of freezing of adsorption water and conversion of the genome into the form of anhydrides. The appearance of a signal with a mass of 130,378,654 Da from

ANC in the loose conformation (**III**) corresponds to the loss of 1.9 water molecules by one pair of repeating bases (a total of 214,233 bp in the genome). Its agreement with the calculated value was no worse than 7.1E-4%. The new, waterless oscillator has a force constant of 6.39112655E-12 N/m. This means that the ring of the active genome begins to collapse already at the stage of exposure for 24 hours at 233 K and turns into anhydride forms with different contents of adsorbed water.

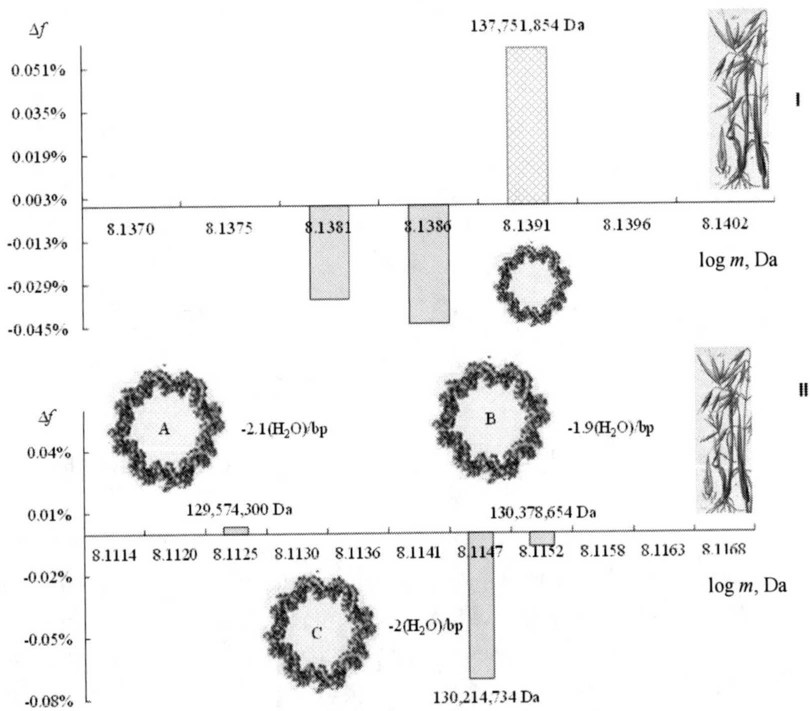

Figure 3. Fragments of the GMS spectra of the oat cytoskeleton (grain, https://ru.wikipedia.org/wiki/Oats) in the region of oscillations of the hydrated (**I**, freshly cut plant) ring of the Ti-plasmid DNA double helix and the dehydrated (**II**) plant. A, B, and C are dehydrogenated rings of the genome of R *radiobacter* strain R58. 07/05/2023.

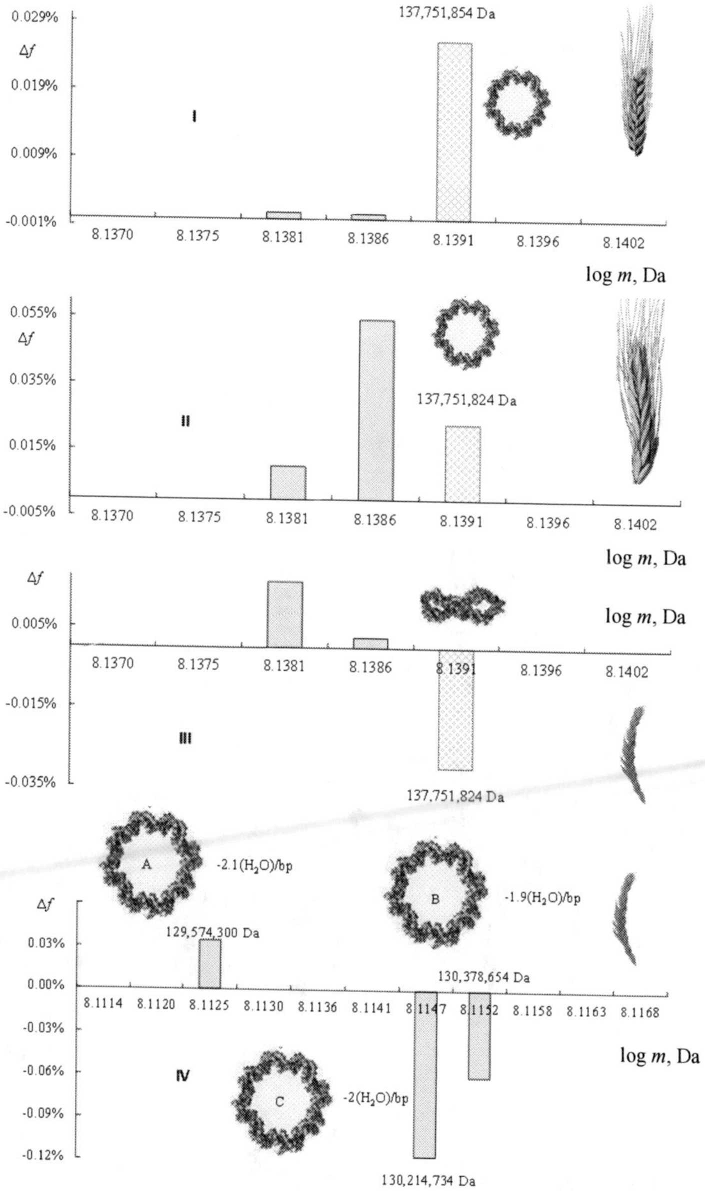

Figure 4. Fragments of GMS spectra of the cytoskeleton of fresh grains of wheat (**I**, https://ru.wikipedia.org/wiki/Wheat), rye (**II**, https://ru.wikipedia.org/wiki/Rye), 04.07.2023 and wild rye (**III** and **IV**, http://agroatlas.ru/ru/content/related/Secale_sylvestre/) in the region of oscillations of hydrated (**I…III**) Ti-plasmid and plasmid in the form of anhydrides in wild rye grains (**IV**), 07.07 .2023.

To destroy anhydrides, broadband gravitational noise (GN) was used, which irradiated wild rye for 60 s. After the irradiation procedure, no signals of either the anhydrides of the genome of R *radiobacter* strain R58 or its hydrogenated form were detected in the sample, Figure 6. At the same time, the activity of tubulins was preserved ($\Delta f = 0.07\%$), the grain did not die.

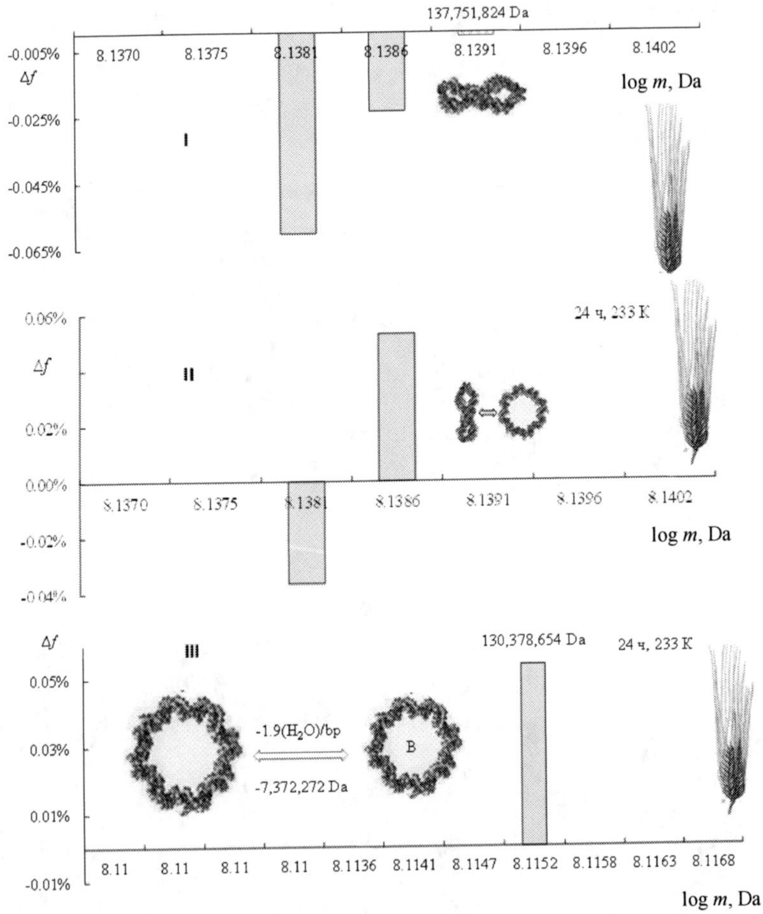

Figure 5. Fragments of the GMS spectra of ANC in the cytoskeletons of fresh barley grains **I**, after 24 hours of exposure at 233 K (**II**), the same, but in the region of the appearance of a signal from oscillations of the dehydrated genome ring with a mass (**III**). 07/04/2023.

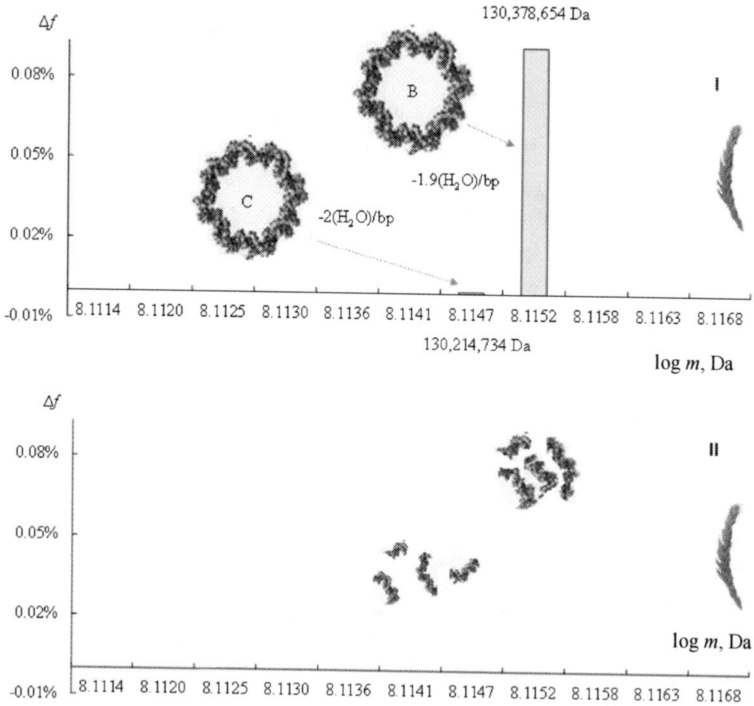

Figure 6. Fragments of the GMS spectrum of anhydrides in the genome of R *radiobacter* strain R58 in wild rye grain before (**I**) and after exposure to GN (**II**). Commentary in the text.

Table 1. The state of tubulin conformations as an energy characteristic of their interaction with their surroundings (Δ*f*). Grains of barley

Barley	Δ*f*	**Condition**
Original grain	0.130%	live
24 h, 233 K	0.193%	live
48 h, 233 K	-0.016%	died
96 h, 233 K	-0.093%	died
40 days, 293 K	-0.006	died

Table 1 presents data on the study of the state of tubulins (https://ru.wikipedia.org/wiki/Tubulin) in barley at critical processing and storage temperatures. The state of tubulin conformations is an indicator of the life or death of a plant [3]. Low values of Δf indicate that the interaction of tubulins with their surroundings through hairpins (https://ru.wikipedia.org/wiki/Hairpin_(biology)) is small, that they are dominated by intra tubulin (coil) interactions, that the division process cells is no longer possible, the plant is dead.

On the contrary, high, positive values of Δf indicate the active interaction of tubulins with their surroundings; the plant is alive. From Table 1 it follows that low-temperature shock destroys the interaction of tubulins with its surroundings and the plant dies. On the other hand, a strong signal at 24 hours indicates a correct understanding of the dynamics of destruction of the Ti-plasmid rings (Figure 4 **B**), according to which, at 24 hours, in the grain the plasmid ring is in a critically tense equilibrium of oscillations of dense and loose conformations preceding decay.

In general, it was observed that the presence of Ti-plasmid in plant cytoskeletons is highly dependent on humidity. Removal of water leads to the disappearance of the genome signal of R *radiobacter* strain R58. This means that the stability of the DNA ring also depends on the water content. The latter stabilizes the conformation of this double helix in a metastable state. It follows that freezing water or simple drying under normal conditions are the best means of preventing cancer in plants.

As can be seen from Table 1, tubulins in all samples, when the grains are kept at low temperatures, are in a state of slow death ($-\Delta f$).

Since the state of energy filling of individual ANC (including Ti-plasmids) depends on the energy state of the entire ensemble of ANC, it makes sense to get an idea of the dynamics of such filling. Table 2 shows the main characteristics of LRO in barley grain, from which it follows that low temperatures reduce the total number of ANCs in the ensemble. At the same time, the part of dense, rich in potential energy ANCs increases. The minimum value of D_c at 48 hours can be interpreted as thermal destruction of the cytoskeleton with subsequent redistribution of energy in a new LRO, with a new thermodynamic equilibrium between all participants in the ANC ensemble. Heating the grain to 293 K leads to the filling of the ensemble with energy and the part of dense ANCs increases to 76%. The readings of the average mass of the ANC (M_{GMS}) in the ensemble change somewhat differently. At the 48th hour of the experiment, with a drop in the number of ANC kind and a decrease in the parts of their dense forms, the M_{GMS} values reach a maximum. This phenomenon can be understood as the thermal destruction of organized (low entropy) structures in the cytoskeleton and the formation of ANC agglomerates with high entropy (destruction of organized structures in the grain cytoskeleton) and stronger, freer oscillations (Δf).

Figure 7 shows fragments of the GMS spectra of Ti-plasmid anhydrides and water clusters in the cytoplasm of potato cells. If the GMS spectrum of the hydrated Ti-plasmid does not contain its signals, then its anhydrides (129,574,300 Da and 130,378,654 Da), in dense, inactive conformations, are reliably present in potato cells. In this case, the basic water cluster $(H_2O)_{12}$ [4, 5] and the $(H_2O)_{25}$ cluster [6, 7] are in loose, active forms. Their conformations are dominated by interactions with their surroundings. The presence of anhydrides in the cytoskeleton of a vegetable indicates the absence of free water molecules in it. All water molecules in potatoes are involved in adsorption, cluster and inclusion forms, in potato tissue.

Table 2. General characteristics of the ANC ensemble (from 200 Da to 3.6 billion Da) in barley grain during its exposure at 233 K and subsequent 40-day storage at 293 K

Barley	N	M_{GMS}, Da	D_c
Start	2533	578,723,266	36%
24 h	2494	592,944,520	54%
48 h	2461	677,220,582	48%
96 h	2495	636,392,813	55%
40 days	2550	507,637,465	76%

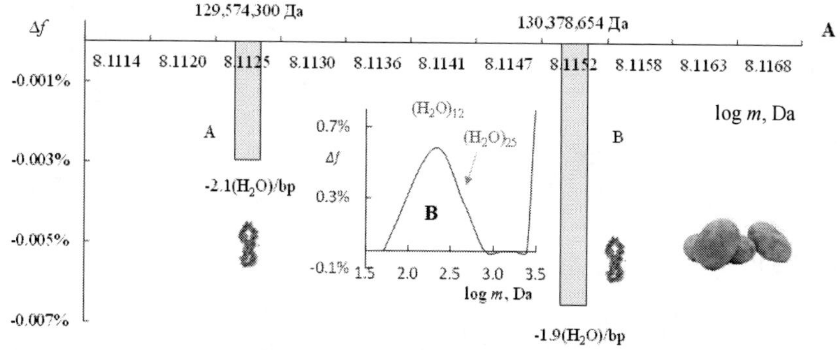

Figure 7. Fragments of GMS spectra: **A** - early potatoes (Belana, [8]), **B** - water clusters. 08/10/2023.

An interesting situation arises with onions (https://ru.wikipedia.org/wiki/Onions_onions). In the central part of the onion, no signals of either the Ti-plasmid or its anhydrides were detected. At the same time, only anhydride signals were recorded in the root part, Figure 8. This should be understood as the beginning of the process of bacterial expansion into the cytoskeleton. Moreover, in the root part of the conformation, two types of anhydrides (A and B) are presented in a dense, inactive form. Both small water clusters are dominated by loose conformations that actively interact with their surroundings. This indicates the youth and freshness of the vegetable. From all of the above it also follows that there are no free water molecules in the cytoplasm of onion cells.

Let us consider the state of the Ti-plasmid in the cytoplasm of animals in the region. Figure 9 shows fragments of the GMS spectrum of the leg tissue of a large garden snail. The spectrum revealed signals of hydrated Ti plasmid (137,891,629 Da), but signals of anhydrides of this plasmid were not recorded. At the same time, loose structures dominated in both small water clusters. Therefore, there is no water deficiency in the snail foot cytoskeleton. Apparently, the hydrogel has the properties of holding water clusters in active conformations and having a sufficient number of free water molecules. Therefore, Ti-plasmids cannot be dehydrated to anhydride forms. Note that the Ti-plasmid has a slightly larger mass than its counterpart in cereals.

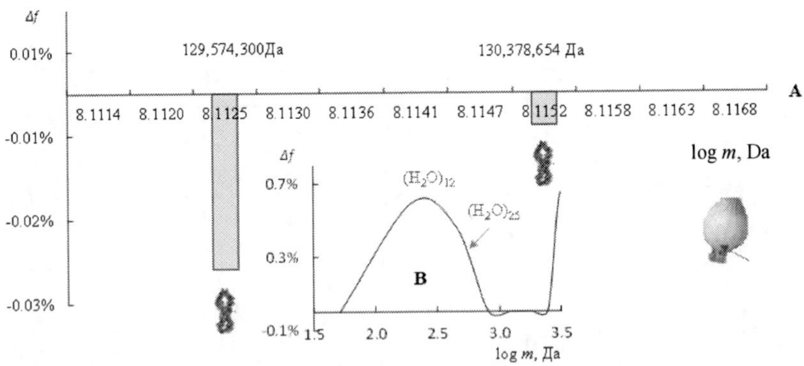

Figure 8. Fragments of GMS spectra: **A** – fresh onions, **B** – water clusters. 08/11/2023.

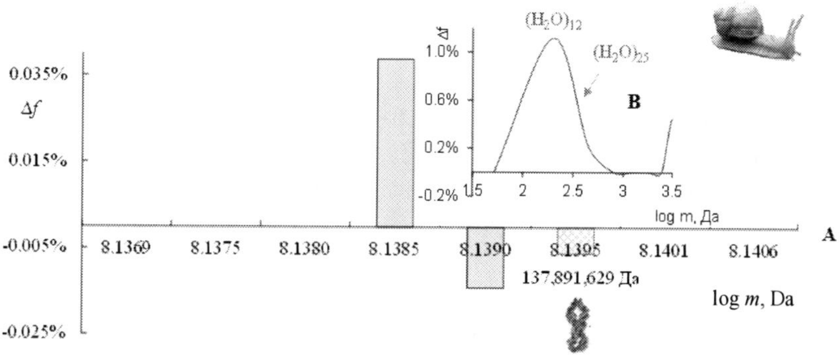

Figure 9. Fragments of GMS spectra: **A** - legs of a large garden snail (*Helix pomatia*), **B** - water clusters. 07/08/2021.

In the snail leg, there are ~10 bp per basic cluster of water $(H_2O)_{12}$ in Ti-plasmid (137,891,629 Da, dominance of dense ring conformations). It should be noted here that the mass of the Ti-plasmid in the cytoplasm of snail cells is greater than the mass of the Ti-plasmid in plants (Figures 1, 3 and 4) by 139,805 Da, which corresponds to 7.761 water molecules. That is, by 100 bp there is an average of one cluster of water. Ti-plasmid anhydrides were not detected in the cytoplasm of snail stalk cells. Since the Ti-plasmid is in a dense conformation, the water clusters in it act as a binder, stabilizing the twisted, dense shape. So, water in the Ti-plasmid is integrated into the ANC as a single oscillator.

In woodlice muscle, no signals of hydrated Ti-plasmid were detected, but in the GMS spectrum there is only one signal from anhydride structures in loose forms with a mass of 130,510,976 Da. The following Figure 10 shows fragments of the GMS spectrum in the region of oscillations of only one anhydride with a mass of 130,510,976 Da in a loose ring conformation. This anhydride differs from its analogues in cereal crops in mass (Figures 2 and 3). This also applies to the hydrated Ti-plasmid, both in the garden snail (137,891,629 Da, Figure 9) and, most likely, in the woodlice, a signal that was not detected in the GMS spectrum. From the GMS spectrum of water it is clear that the signals of the clusters are very weak and reflect only the weak dominance of dense conformations in them, with poor interaction with their surroundings. In woodlice muscle cells, on average, by 133 bp there is one water cluster $(H_2O)_{25}$ (the signal from the base water cluster is neglected here). It can be assumed that this cluster is in the form of inclusive (captured) water in the anhydride rings of the Ti-plasmid. It is integrated into a single oscillator with the anhydride. Let us remember that woodlice live in predominantly humid conditions. In our case, woodlice lived in a relatively dry and cold basement (7 ± 3°C).

It should be noted that no signals from the Ti-plasmid or its anhydrides were detected in the muscles of the grasshopper (*Schistocerca americana*), frog (*Pelophylax esculentus*), and earthworm (*Lumbricidae*).

The following Figure 11 shows fragments of the GMS spectra of signals from anhydride and water clusters in the heart muscle of a domestic chicken. As can be seen, water clusters are represented by loose conformations with strong interaction with the surroundings. At the same time, no Ti-plasmid signal was detected, but there was an oscillation signal from one of its anhydrides in a dense conformation, with a mass of 130,510,976 Da. Note that the mass of this anhydride is slightly greater than the mass of analogues in cereals, potatoes and other studied samples (130,378,654 Da). Such events can be understood as the presence of water clusters in the oscillating ring of the anhydride, as a single binder of ANC, in the form of included water.

It is reasonable to believe that these water clusters are also involved in interactions with other actors in the cytoplasm of cells. In heart muscle cells, on average 133 bp there is one averaged water cluster: $(H_2O)_{12}$ and $(H_2O)_{25}$.

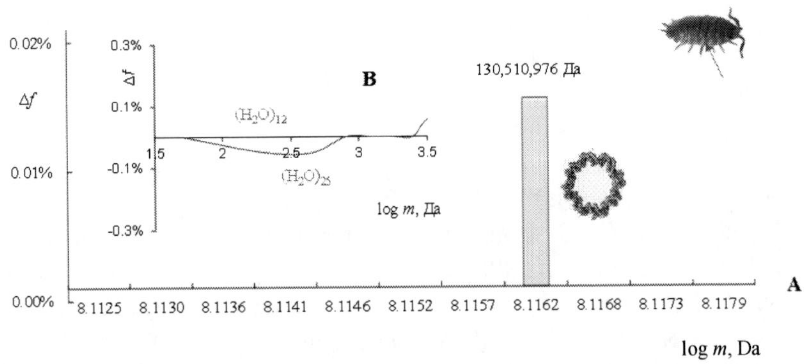

Figure 10. Fragments of GMS spectra: **A** - muscles of the most ancient representative of eukaryotes - woodlice (https://ru.wikipedia.org/wiki/Woodlice, (*Porcellio scaber*)), the arrow shows the location of the sensor, **B** - water clusters. 07/18/2021.

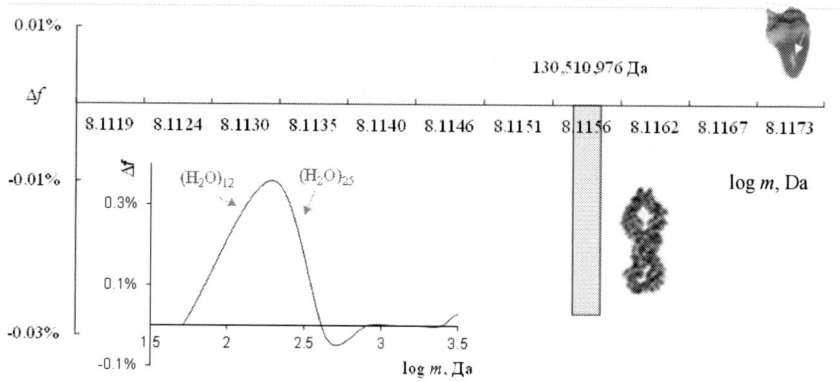

Figure 11. Fragments of GMS spectra of fresh heart muscle of domestic chicken (*Gallus gallus domesticus*), the arrow shows the location of the sensor. 07/13/2021.

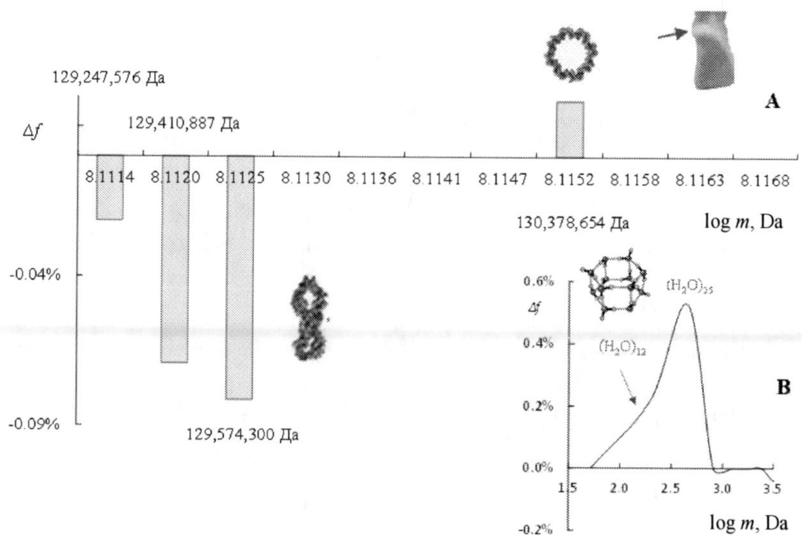

Figure 12. Fragments of GMS spectra in the region of oscillations: **A** - Ti-plasmid anhydrides in the cytoplasm of skin cells (epidermis and dermis) of the human heel and **B** - basic water cluster. 05/17/2023.

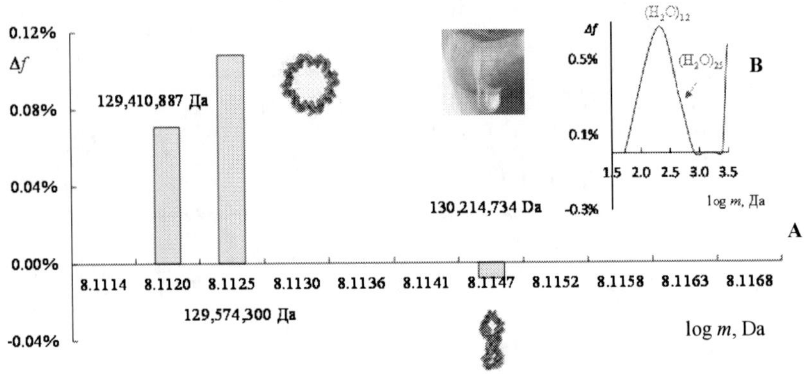

Figure 13. Fragments of GMS spectra in the region of oscillations: **A** - anhydrides of Ti-plasmids in probant-48 saliva, **B** - small water clusters. 01/19/2024.

The following Figure 12 presents experimental data on the analysis of Ti-plasmid anhydrides in the dermis of the skin of a probant living in northern Germany for more than 20 years. GMS spectroscopy did not detect signals of the hydrated Ti-plasmid (137.7 MDa), but immediately 4 types of its anhydride states in different conformations (**A**). From spectrum **B** it is clear that in the dermis is dominated by water clusters of 25 molecules in loose conformations. That is, water clusters actively interact with their surroundings. Therefore, it can be assumed that the loose anhydride plasmid, with a mass of 130,378,654 Da, is involved in such interactions. He is active and can pose a danger to his surroundings.

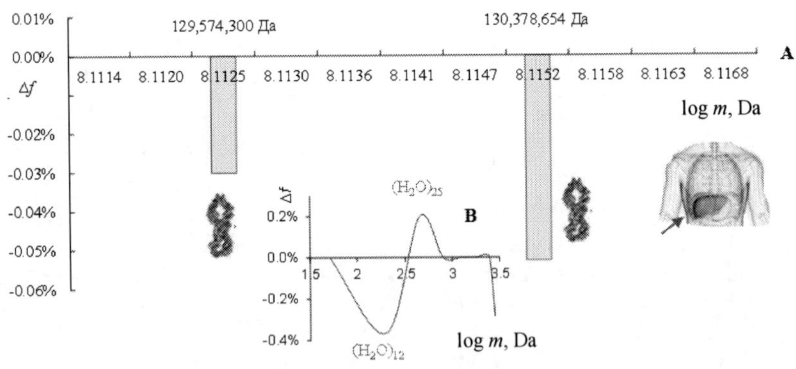

Figure 14. Fragments of GMS spectra in the region of oscillations: **A** - Ti-plasmid anhydrides in the liver of probant-70 (non-invasive scanning of GN in the liver area), **B** - small water clusters. 01/26/2022.

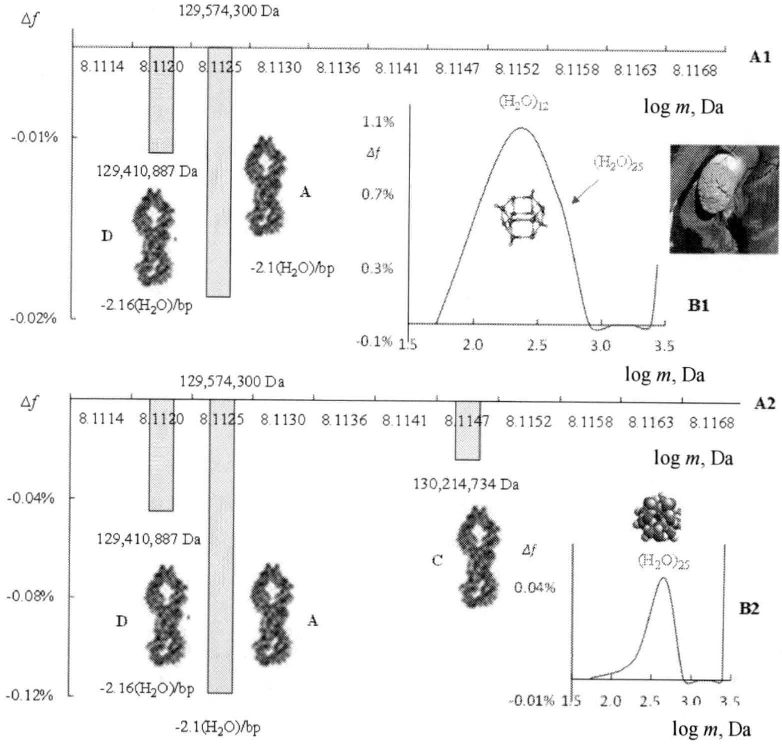

Figure 15. Fragments of GMS spectra in the region of oscillations: **A1** - anhydrides of Ti-plasmids and water (**B1**) in micro biocenoses of liquid cow manure (farm 1). 08/22/2023 [9]. **A2** and **B2**, the same, but for farm 2 [10]. 10/12/2023. The water cluster model is taken from [6].

Figure 13 shows two GMS spectra of oscillations of Ti-plasmid anhydrides and water clusters in the saliva micro biocenosis of probant-48 (48 years old). It can be seen that two types of anhydrides are in loose conformations (129,410,887 Da and 129,574,300 Da) and one in which dense forms dominate (130,214,734 Da). At the same time, basic water clusters in loose conformations also dominate in saliva. Thus, from the above, we can conclude that there are no free water molecules in saliva and its biocenosis. All water is presented in the form of adsorption, included and cluster forms.

Figure 14 shows fragments of the GMS spectra of Ti-plasmid anhydrides and water clusters in the ensemble of molecular oscillators of the liver Probant-70 (70 years old). Here, in both anhydrides, dense forms dominate, inactive states. In the region of oscillations of water clusters, the situation is unusual. Dense forms dominate in the base cluster, while loose forms dominate in the larger cluster. Moreover, the energy contribution (abs(Δf)) to the large ensemble of oscillators of the base cluster is almost twice as large as that of a cluster of 25 molecules. But the $(H_2O)_{25}$ cluster is active, it interacts with its surroundings, but the base one does not. Thus, there are also no free water molecules in the cytoplasm of liver cells. All water is involved in adsorption interactions with all molecular actors of the cells of this organ.

Figure 15 shows fragments of GMS spectra in the region of water oscillations and Ti-plasmid in micro biocenoses of liquid cattle manure (two cow farms). No Ti-plasmid signals were detected in the manure of cows from both farms, but signals from dense conformations of its anhydrides were recorded, similar the masses of anhydrides D and A, Figure 3. Relative to the hydrated Ti-plasmid (137,751,824 Da), it, in the form of anhydrides, lost 2.6(H_2O) and 2.1(H_2O) molecules per repeat bp, respectively. The difference between spectrum **A1** and **A2** is detected in the presence of signals from another anhydride C (130,214,734 Da), which we explain by the lower water content in manure (spectra **B1** and **B2**).

The quiet expansion of R *radiobacter* in northern Germany occurs as a result of disruption of the biocenosis due to intensive agriculture using liquid cattle manure as fertilizer, Figure 16.

The danger of expansion, the entry of the Ti-plasmid into resource-rich old, sick, damaged or weakened eukaryotic cells, we see in the fact that the plasmid, like the brain of a bacterium, will be engaged in the regeneration of the entire lost infrastructure of the bacterial cell (*endoplasmic reticulum*) through the specific absorption of resources. In this case, obligatory gene mutations will arise with unknown consequences.

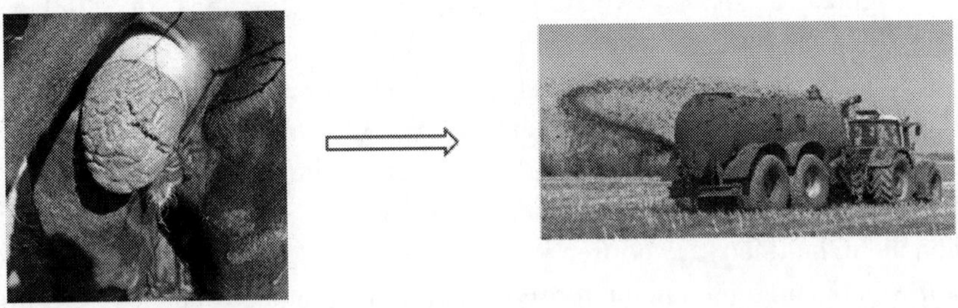

A B

Figure 16. Main route of spread of Ti plasmid into the environment. **A** – emission of manure by cattle, **B** – splashing of manure on agricultural land.

Conclusion

The reason for the expansion of R *radiobacter* strain R58 *via* a Ti plasmid in northern Germany is the use of cattle manure as a fertilizer for growing cereal crops.

Cereal crops fertilized with manure are a source of infection of eukaryotes, including humans [9].

Eukaryotes that do not consume grain feed are not infected with Ti plasmids.

Under conditions of water deficiency, Ti-plasmids are easily converted into their anhydride forms.

Ti-plasmid anhydrides are widely present in the cytoskeletons of eukaryotic cells and, most likely, in the cytoplasm of the cells themselves.

In eukaryotic cells infected with Ti plasmids, there are no free water molecules. The cytoplasm of cells, therefore, experiences an acute shortage of water.

Water in eukaryotic cells is represented by small clusters: $(H_2O)_{12}$ and $(H_2O)_{25}$.

In the micro biocenosis of liquid manure, Ti-plasmids persist for a long time and, once in the soil, are able to diffuse into the stem and further into the grain along with irrigation water [11].

The expansion of R *radiobacter* strain R58 through the Ti plasmid in northern Germany through the food chain poses a serious threat to the population and requires more in-depth research into this danger.

The superposition of regional gravitational noise with the noise of the Universe forces the formation of ANC in cereal grains typical of northern Germany.

The state of tubulin conformations makes it possible to quickly determine the viability of a plant after exposure to drought and low and high temperatures.

In freshly harvested cereals, the genome of R *radiobacter* strain R58 is presented in the form of rings of DNA double helices with adsorption layers of water of three types: 2.2; 2.1 and 1.9 water molecules per DNA repeat base.

Removal of water (drying or freezing at 233 K) leads to the formation of anhydrides of the genome of R *radiobacter* strain R58 with different conformations of DNA rings.

Wheat cultivars did not reveal signals from the genome of R radiobacter strain R58 and its anhydrides.

A method for destroying the genome of R *radiobacter* strain R58 has been proposed.

Sponsoring

To continue work in this direction, we are looking for sponsors. Write to us.

References

[1] Zubow K., Zubow A. V., Zubow V. A. The Way to the ETIs. *Applied gravitational mass spectroscopy.* Nova Sci. Publ. NY, 2014. https://www.novapublishers.com/catalog/product_info.php?products_id=42668&osCsid=5bd85d42dc273360fd48126de7be9daf.

[2] https://www.sciencedirect.com/topics/medicine-and-dentistry/ti-plasmid-plant.

[3] Zubow K. V., Zubow A. V., Zubow V. A. The ensemble of tubulins in onion roots and their properties. *Russian agricultural science*. 2018. 6: 21-25. *Russian Agricultural Sciences.* Allerton Press, Inc. 2019, 45: 30-34. ISSN 1068-3674. Published in Russian in Rossiiskaya Sel'skokhozyaistvennaya Nauka, 2018, 6: 21–25. doi: https://doi.org/10.3103/S1068367419010221.

[4] Lenz A., Lars Ojamäe. (2006). On the stability of dence versus cage-shaped water clusters: Quantum-chemical investigations of zero-point energies, free energies, basis-set effects and IR spectra of $(H_2O)_{12}$ and $(H_2O)_{20}$. *Chem. Phys. Letters*. 418: 361-367.

[5] Bogdanov E. V., Manturova G. M. (2000). Equimolecular model of water. *Biomedical Electronics*. 2000. № 7, 19-28.

[6] https://www.google.com/search?sca_esv=602110905&rlz=1C1CHBD_deDE858DE858&q=water+clusters+of+25+molecule+radius&tbm=isch&source=univ&fir=gxDhe07-nG-x-M%252CSNRsXQvBcfRfvM%252C_%253Bj_bgz0AZkFO-cM%252CEyRQxz9EXIYhIM%252C_%253B3fWCulN2-pj3oM%252CTBA_U5dCXXu4eM%252C_%253BG1EDB6Q43iBjTM%252CUbt2cPz8sqBolM%252C_%253BlYfnGNbsSQvseM%252CErXMnNrSQQlE8M%252C_%253BgkNQJCsVVcqtuM%252Ci7rYcNs9wwBK9M%252C_%253BU5qqrbUtbp4jhM%252CUbt2cPz8sqBolM%252C_%253BLjxhmRIx-OGkdM%252Cc3x-WxjiyQ7dyM%252C_%253BiLKyl_QJtwp4JM%252CvRGgi84QswFhiM%252C_%253BEVTukF3-ApmhoM%252CxzCoN57-xMc0PM%252C_&usg=AI4_-kRwFppwTUYzaOXgMgfFb9J5QSVChA&biw=1707&bih=731&dpr=0.8#imgrc=s2tk4tADnNeaXM.

[7] Chaplin M. *SBU London*, http://www.lsbu.ac.uk/water/index.

[8] https://de.wikipedia.org/wiki/Belana_(Kartoffel)#:~:text=Als%20festkochende%20Kartoffel%20ist%20Belana,aufgrund%20hoher%20Keimruhe%20lange%20lagerf%C3%A4hig.

[9] Zubow A., Zubow K., Zubow V. A. Dynamics of genomes in a liquid cow manure in the food traffic. *Agricultural Research Updates*. 2023. Volume 47. Ed. Charles D. Grant. Nova Publisher Inc. NY. 2024, in print.

[10] Zubow A., Zubow K., Zubow V. Systematicity in the construction of mass clumps in microbiology. Clusters, domains, coils, genes, genomes, plasmids, nucleosomes. Horizons in World Physics. Ed. Albert Reimer. NY, 2024, vol. 312, pp. ... in print. *Horizons in World Physics*. Volume 312 – Nova Science Publishers (www.novapublishers.com).

Chapter 2

Effects of Methyl Tertiary-Butyl Ether (MTBE) on Testis Tissue and Sex Ratio of Offspring in Adult Rats

Leila Khalili[1,*], PhD
and Abbas Khalili[2], MSc

[1]Department of Basic Sciences, School of Veterinary Medicine, Shiraz University, Shiraz, Iran
[2]Faculty of Natural Resources and Marine Sciences, Tarbiat Modares University, Tehrn, Iran

Abstract

Methyl tertiary-butyl ether (MTBE) is used to reduce carbon monoxide and ozone in urban air and to boost fuel octane. Also, it is an oxygenated fuel additive which has been used widely in many parts of the world. MTBE can migrate faster and farther in the ground than other components of gasoline. This makes it more likely to contaminate public water systems and people could be exposed to MTBE by drinking, swimming, or showering in water contaminated with MTBE. Based on several studies on carcinogenicity of MTBE in animals, it has been shown that chronic exposure to MTBE is associated with renal tubular cell tumors, hepatocellular adenomas and Leydig cell tumors of the testes in rodents.

We studied the effect of oral exposure to MTBE on histologic and histomorphometric changes of testis in adult rats. We also investigated the effect of MTBE on offspring sex ratio, sex hormones and antioxidant enzyme. A total of 25 adult male Sprague-Dawley rats were randomly

---

[*] Corresponding Author's Email: khalelili2007@yahoo.com.

In: Advances in Biology. Volume 6
Editor: Charles D. Grant
ISBN: 979-8-89113-712-7

divided into five equal experimental groups which received 400, 800, and 1600 mg/kg/day MTBE in almond oil by gavages for 30 consecutive days.
At the end of the experiment, blood samples were taken for determination of sex hormones and antioxidant enzymes. Then, male rats were mated with healthy unexposed female rats and sex of offspring was determined after birth. Histomorphometric analysis showed no significant difference in absolute and relative testis weight, connective tissue thickness, germinal epithelium height, tunica albuginea thickness and Sertoli cell numbers between experimental groups ($P > 0.05$). However, trend analysis showed that the seminiferous tubule diameter increased and interstitial cell numbers as well as spermatocyte and spermatid cell numbers decreased significantly in MTBE treated groups ($P < 0.05$). It may be concluded that MTBE could exert adverse effects on spermatogenic cells in adult rats. There was a significant decreasing trend for luteinizing hormone (LH) and testosterone in experimental groups (rs = -0.50, P = 0.030 and rs = -0.67, P = 0.002, respectively). No changes were observed for superoxide dismutase. However, decrease in glutathione peroxidase (GPX) was observed in all treatment groups compared with control which was significant in 400 mg/kg/day MTBE group (P = 0.016). The present study showed that paternal exposure to oral MTBE has no effect on offspring sex ratio; while MTBE exposure could exert dose-dependent changes in serum testosterone and LH in treatment groups. Whether the observed changes in the present study are due to the direct effect of MTBE via passing blood-testis barrier or its indirect effect through another mechanism should be elucidated in future studies and need to be clarified.

Keywords: methyl tertiary butyl ether (MTBE), sex ratio, sex hormones, histomorphometric, testis

Introduction

Methyl tertiary-butyl ether (MTBE) is the main component of oxygenated fuel which has been used worldwide to reduce air pollution, increase the oxygen content of gasoline and decrease carbon monoxide (Stern and Kneiss, 1997). MTBE can penetrate into the ground and contaminate public water systems. Therefore, people could be exposed to MTBE by swimming, or drinking contaminated water. Based on several studies on carcinogenicity of MTBE in animals, it has been shown that chronic exposure to MTBE is associated with renal tubular cell tumors, hepatocellular adenomas and Leydig cell tumors of the testes in rodents (Bird et al., 1997; Clegg et al., 1997; Belpoggi et al., 1995).

MTBE has been listed as a potential human carcinogen (USEPA, 1998). It has been shown to induce interstitial cell tumors of the testes in rats in inhalation and in gavage studies (Belpoggi et al., 1995; Bird et al., 1997; Clegg et al., 1997). In addition to MTBE carcinogenicity, there are other published studies of potential reproductive and endocrine effects of MTBE in male rodents (Day et al., 1998, 2000; Billiti et al., 2005; Li et al., 2008; Bermudez et al., 2012). The results of several toxicity studies suggest that testis is one of the primary target organs in rodents (Williams, 2000; Li et al., 2007, 2008, 2009). Despite the testis being a potential target organ for MTBE toxicity, little is known about MTBE-induced histopathological changes in testis. Li et al. (2008) reported an alteration in the testis of high dose MTBE treated groups (800 and 1600 mg/kg/day for two and four weeks), including an irregular and disordered arrangement of cells in the seminiferous tubules and shedding of cellular material in the seminiferous epithelium in rats. On the other hand, histomorphometric examination of tissues has a prominent role in male reproductive toxicity evaluation. It could provide information on the severity of toxicity and cellular site of the damage. Based on the results of carcinogenicity studies in male and female rats, it has been suggested that MTBE might interfere with normal hormonal feedback mechanisms (Williams et al., 2000). In contrast to MTBE carcinogenicity, there are a few recently published studies of potential endocrine effects of MTBE, and associated male hormonal effects have not been addressed in detail. Furthermore, the studies on male reproductive hormones had inconsistent results. For example, an increase, a decrease and no changes in testosterone levels have been reported in previous studies (Day et al., 1998; Williams et al., 2000; Billiti et al., 2005; Li et al., 2008; Bermudez et al., 2012). On the other hand, one interesting aspect of reproduction is the sex ratio of off-spring at birth. Considering the growing body of knowledge about the association of offspring sex ratio with many factors including environmental exposures (Rosenfeld and Roberts, 2004; Mocarelli et al., 2000; James, 1996), there is no study concerning the effects of MTBE exposure in male on the offspring sex ratio at birth. It has been shown that paternal exposure to gasoline (containing MTBE) in filling station workers was associated with decrease in offspring sex ratio (Ansari-Lari et al., 2004). Also, sex ratio was skewed to female births in Sprague-Dawley rats exposed to gasoline vapor for 6 hours for 30 consecutive days (Ansari-Lari and Tanideh, 2009). However, to the best knowledge of the authors, studies which have addressed the association of

MTBE exposure with offspring sex ratio were conducted either on female animals after conception and during gestation (Conaway et al., 1985; Bevan et al., 1997), or on both male and female animals prior to mating with continuous exposure of females during pregnancy (Biles et al., 1987).

Furthermore, there are very limited data about the role of oxidative stress in reproductive effect of MTBE (Li et al., 2007, 2008, 2009). It has been shown that MTBE increased the production of reactive oxygen species (ROS) and enhanced lipid peroxidetion in isolated rat spermatogenic cells (Li et al., 2007), can exert reproductive system toxicity of male rats and disturb the secretions of sex hormones possibly due to induction of oxidative stress (Li et al., 2008) and also has a direct toxic effect on cultured rat Sertoli cells (Li et al., 2009).

Taken together, the objective of the present study was to determine the effect of oral administration of MTBE on offspring sex ratio at birth, changes in sex hormones and two antioxidant enzymes, superoxide dismutase (SOD) and glutathione peroxidase in adult male Sprague-Dawley rats. To the best of the authors' knowledge, there is no study concerning the histomorphometric evaluation of testis following exposure to MTBE. Therefore, the objective of the study was to determine the effect of oral administration of MTBE on histological and histomorphometric changes of testis and sex ratio in adult Sprague-Dawley rats. for this aim a total of 25 adult Sprague-Dawley were randomly divided into five equal experimental groups: control, almond oil and three treatment groups which received 400, 800 and 1600 mg/kg/day MTBE in almond oil by gavages for 30 consecutive days. At the end of the exposure period, male rats were anesthetized with ether, and blood samples were individually obtained from heart. Then, animals in all experimental groups were mated with 44 healthy unexposed female rats and then gender of offspring was determined after birth. Then their testes were removed, and after processing for paraffin embedding, used for histological and histomorphometric studies at the light microscopic level.

Seminiferous tubule diameter, germinal epithelium height, the connective tissue thickness and tunica albuginea thickness were measured using the Axiovision L.E. 4.5 software. For measuring the number of Interstitial cells, Sertoli cells and spermatogenic cells, Image tools 3 software was used. Jonckheere's analysis was used to determine their dose-response trend. Data were presented as mean, median and standard deviation (SD). Sex hormones and antioxidant enzymes activities as well as litter size were compared between groups using nonparametric analysis of variance (Kruskal-Wallis tests) followed by Mann-Whitney U tests; Spearman's rho correlation coefficient was used to determine their dose response relationship. Offspring sex ratio at birth (expressed as male proportions) was determined for individual male rats in each group. After arc-sinus transformation, mean of group sex ratio was compared using one-way analysis of variance. Fertility index was calculated in two ways: as number of fertile males to all males and as number of pregnant females to all mated females in each group; statistical comparison between experimental groups was conducted using Chi-square for trend analysis. Data were analyzed using SPSS statistical soft-ware (Version 16.0; SPSS, Inc., Chicago, USA) and a P-value less than 0.05 was considered statistically significant in all analyses.

Study of Histologic and Histomorphometric Changes of Testis

Mean body weight was increased from the start of the study toward the end of the experiment ($P < 0.001$) with no statistical difference between experimental groups ($P = 0.40$). Average food consumption was 169 ± 32 g in the first week and increased to $349 + 28$ in the last week of the study ($P < 0.001$). No significant difference was detected for absolute ($P = 0.75$) and relative ($P = 0.28$) testis weight (g) between experimental groups (Table 1). Histological studies showed no lesions in control and almond groups (Figs. 1a-b). In all MTBE treatment groups, changes were observed in seminiferous tubule as pyknosis, decreased cell layers, increased cell distance as well as the presence of vacuoles in the epithelium (Figs. 2a-c). The severity of lesions enhanced with increased dose of MTBE. The results of the statistical comparison of histomorphometric variables are presented in Tables 2 and 3. No significant changes were observed for connective tissue thickness ($P = 0.48$), germinal epithelium height ($P = 0.32$) and tunica albuginea thickness ($P = 0.95$) in controls compared with MTBE treated groups. However, seminiferous tubule diameter showed significant increase in MTBE treated groups compared with almond oil group ($P = 0.023$, Table 2). On the other hand, results showed a decrease in interstitial cell numbers as well as spermatocyte and spermatid cell numbers in MTBE treated groups in comparison with control ones ($P < 0.05$). Spermatogonia ($P = 0.15$) and Sertoli cell numbers ($P = 0.98$) did not show significant changes (Table 3).

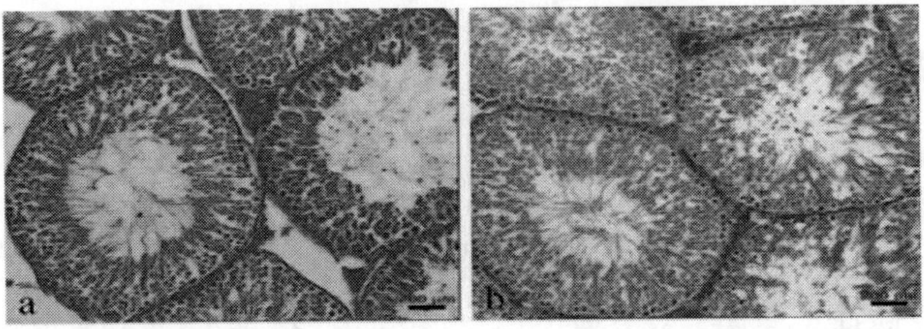

Figure 1. Histological section of adult rat testis in experimental groups. a: Control, and b: Almond oil (scale bar 200 μm, H&E).

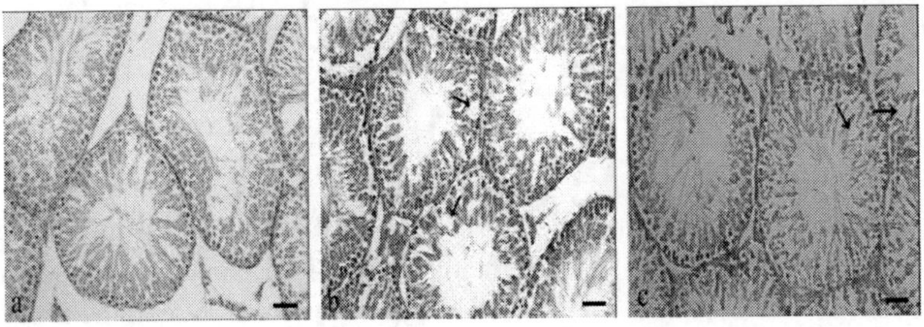

Figure 2. Histological sections of adult rat testis in treatment groups. a: 400, b: 800, and c: 1600 mg/kg/day MTBE. In Figs. 2b and c, arrowheads indicate presence of vacuoles in epithelium (scale bar 200 μm, H&E).

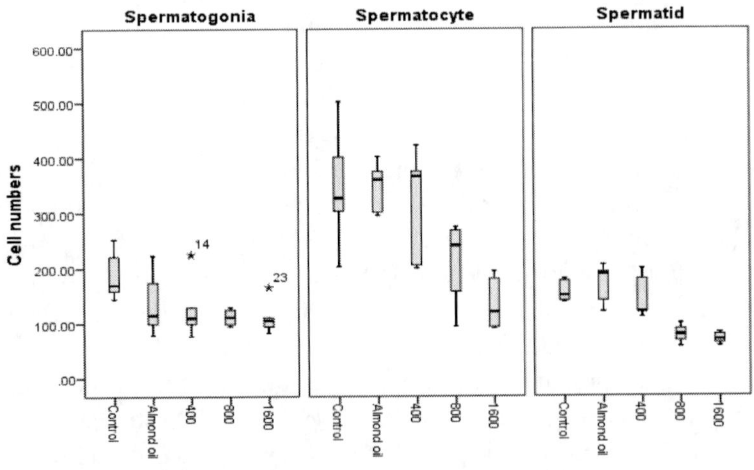

Figure 3. Box plots showing minimum, maximum and quartiles of number of cells of testis in controls and MTBE (mg/kg/day) treated groups in adult Sprague-Dawley rats. Asterisks show outliers and extreme values, respectively.

Table 1. Comparison of testis and body weights (gr) in MTBE (mg/kg/day) treated groups with control groups in adult Sprague-Dawley rats

Experimental groups		Absolute testis weight	Relative testis weight	Final body weight
Control	Mean	1.20	0.004	267.40
	Std. Deviation	0.45	0.001	44.27
	Median	1.00	0.004	295.00
Almond oil	Mean	1.60	0.006	271.60
	Std. Deviation	0.89	0.003	33.15
	Median	1.00	0.004	276.00
400mg	Mean	1.40	0.005	303.80
	Std. Deviation	0.55	0.002	5.21
	Median	1.00	0.003	300.00
800 mg	Mean	1.25	0.004	291.75
	Std. Deviation	0.50	0.001	19.77
	Median	1.00	0.004	292.50
1600 mg	Mean	1.60	0.006	272.40
	Std. Deviation	0.55	0.002	25.08
	Median	2.00	0.006	265.00

No statistical difference was observed between groups ($P > 0.05$).

Table 2. Comparison of thickness (T), diameter (D) and height (H) of various parameters in testis of MTBE (mg/kg/day) treated groups with control groups in adult Sprague-Dawley rats

Experimental groups		Tunica albuginea T.	Connective tissuse T.	Seminiferous Tubules D.*	Germinal Epithelum H.
Control	Mean	42.37[a]	46.59[a]	289.91[a,b]	98.49[a]
	Std.Dviation	15.49	8.33	57.62	8.37
	Median	43.09	44.80	284.84	97.95
Almond oil	Mean	42.08[a]	50.86[a]	267.92[a]	104.71[a]
	Std.Deviation	3.12	10.53	21.70	5.03
	Median	42.1500	46.94	263.14	105.47
400mg	Mean	41.77[a]	53.24[a]	319.78[ab]	110.65[a]
	Std.Deviation	6.57	10.16	29.32	16.29
	Median	39.97	56.20	321.24	109.49
800 mg	Mean	42.62[a]	53.04[a]	342.62[b]	109.83[a]
	Std.Deviation	6.67	13.24	25.99	6.78
	Median	42.74	58.64	341.65	111.63
1600 mg	Mean	41.93[a]	60.86[a]	345.78[b]	110.30[a]
	Std.Deviation	9.44	14.81	38.90	12.01
	Median	37.26	60.19	320.70	109.96

Different superscript letters in each column show significant differences ($P < 0.05$).

* Show significant trend analysis by Jonckheere test ($P < 0.05$).

When trend analysis using Jonckheere's test was performed, results showed that there are significant decreasing trends for number of cells including interstitial (P = 0.002), spermatogonia (P = 0.024), spermatocyte (P = 0.001) and spermatid (P = 0.001) in the experimental groups. For spermatogenic cells, this trend is depicted in Figure 3. The increasing trend Figure 1: Histological section of adult rat testis in experimental groups. a: Control, and b: Almond oil (scale bar 200 μm, H&E).

Table 3. Comparison of number of cells in testis of MTBE (mg/kg/day) treated groups with control groups in adult Sprague-Dawley rats

Experimental groups		Number of cells				
		Sertoli	Interstitial*	Spermatogonia*	Spermatocyte*	Spermatid*
Control	Mean	3.60 [a]	5.00 [a]	189.20 [a]	347.20 [ab]	156.80 [a]
	Std.Deviation	0.55	0.71	45.49	111.82	19.93
	Median	4.00	5.00	170.00	327.00	150.00
Almond oil	Mean	3.60 [a]	4.40 [a]	137.80 [a]	346.60 [a]	169.20 [a]
	Std.Deviation	0.89	1.34	59.52	46.90	36.97
	Median	3.00	5.00	115.00	360.00	189.00
400mg	Mean	3.60 [a]	4.00 [a]	127.80 [a]	313.80 [ab]	145.40 [a]
	Std.Deviation	0.55	1.22	57.62	103.90	40.12
	Median	4.00	4.00	110.00	366.00	120.00
800 mg	Mean	3.50 [a]	3.50 [a]	111.50 [a]	212.25 [bc]	77.75 [b]
	Std.Deviation	0.58	0.58	15.72	82.12	17.56
	Median	3.50	3.50	111.00	240.50	78.00
1600 mg	Mean	3.40 [a]	3.00 [b]	111.20 [a]	135.40 [c]	69.80 [b]
	Std.Deviation	1.14	0.71	31.95	48.78	10.33
	Median	3.00	3.00	105.00	120.00	69.00

Different superscript letters in each column show significant differences (P < 0.05).
*Show significant trend analysis by Jonckheere test (P < 0.05).

Study Sex Ratio, Antioxidant Enzyme Activities and Serum Sex Hormones

Overall, 149 male and 155 female were produced by experimental groups. No still-birth or gross anomaly was detected in experimental groups. Summary statistics for re-productive parameters are displayed in Table 4. No significant difference between groups sex ratio (P = 0.91) or number of pups at birth (P = 0.11) was detected (Table 4). Fertility index based on pregnant females was equal to 81.8% for the first three experimental groups and 54.5% for 1600 mg/kg MTBE group (P = 0.22). When fertility index was calculated based on fertile males, it was 100% for the first three groups and 60% for 1600 mg/kg MTBE group (P = 0.058).

Table 4. Summary statistics for reproductive outcomes in MTBE-treated adult Sprague- Dawley male rats

	MTBE (mg/kg/day)			
Reprocactive parameter	0	400	800	1600
Number of males mated	5	5	4	5
Number of fertile males	5	5	4	3
Fertility index, males (%)[a]	100	100	100	60
Number of females mated	11	11	11	11
Number of pregnant females	9	9	9	6
Fertility index, females (%)[b]	81.8	81.8	81.8	54.5
Number of alive litters	9	9	9	6
Mean number of pups at birth[c] (SD)[d]	10.3(2.4)	8.9(2.1)	8.1(1.5)	10.7(2.1)
Number of females pups	47	39	42	27
Number of males pups	46	41	31	31
Mean of individual male sex ratio (SD)	0.48(0.16)	0.50(0.20)	0.43(0.24)	0.50(0.19)

[a] fertile males to mated males, [b] pregnant females to mated females, [c] for pregnant females, [d] standard deviation.

Table 5. Comparison of serum luteinizing hormone (LH, mlU/ml), follicle stimulating hormone (FSH, mlU/ml), testosterone (ng/ml), superoxide dismutase (SOD,U/Hb) and Glutathione peroxidase (GPX, UHb) in MTBE-treated (mg/kg/day) adult Sprague-Dawley male rats

Parameter	MTBE (mg/kg/day)							
	Mean		Mean		Mean		Mean	
	(SD)[a]	0 Median	(SD)	400 Median	(SD)	800 Median	(SD)	1600 Median
FSH	4.90 (3.3)	4.10	4.50 (4.4)	2.70	2.98 (3.0)	1.80	2.35 (2.3)	1.40
LH**	3.79 (2.7)	3.50	3.38 (2.9)	2.90	0.91 (0.6)*	0.62	1.36 (1.1)	0.97
Testosterone**	11.66 (4.2)	10.2	9.98 (2.7)	10.50	5.50 (2.9)	4.50	4.98 (3.3)*	4.10
SOD	5.82 (1.0)	5.55	6.19 (1.3)	6.22	5.88 (3.9)	7.21	6.35 (0.9)	6.22
GPX	185 (181.8)	117.77	40.32 (15.0)	50.40	88.25 (37.3)	84.05	70.60 (32.3)	67.20

[a] Standard Deviation.

* Significant difference with control group ($p < 0.05$).

** significant dose – response relationship ($p < 0.05$).

Comparisons of sex hormones between study groups are presented in Table 5. There was no significant difference (P = 0.33) between study groups for level of follicle-stimulating hormone (FSH). Luteinizing hormone (LH) was significantly lower in 800 mg/kg/day (P = 0.032), and testosterone was lower in 1600 mg/kg/day MTBE group compared with control one (P = 0.032). Spearman's rho correlation coefficient for levels of serum FSH, LH and testosterone with ordered dose groups were -0.43 (P = 0.064), -0.50 (P = 0.030) and -0.67 (P = 0.002), respectively. These decreasing trends with increasing dose (negative dose-response relationship) a Levels of antioxidant enzyme activities in different experimental groups are shown in Table 5. No changes were observed for SOD in this study. However, a decrease in glutathione peroxidase (GPX) was observed in all treatment groups compared with control one which was significant in 400 mg/kg/day MTBE group (P = 0.016). Furthermore, a tendency for significance (P = 0.09) was detected between 1600 mg/kg/day MTBE and control group (Table 5). No dose-response relationship was detected for antioxidant enzymes.

Conclusion

In the present study using histomorphometric evaluation of testis, we observed that oral administration of MTBE could exert adverse effects on spermatogenesis process and cause some histological changes. Control groups showed a compact and regular arrangement of spermatogenic cells in the seminiferous tubules whereas in all MTBE treatment groups, deceased cell layers and increased cell distance were observed. This agrees with the study of Li et al. (2008) who showed that seminiferous tubules of rats in high dose MTBE treated groups had less impact cells and irregular and disordered cell arrangement. Also, Billiti et al. (2005) evaluated the acute effects of oral MTBE exposure on testicular function in male CD-1 mice. Histopathological examination showed a slight increase in the number of tubules with gross disruption in the 2000 mg/kg group. Decreasing trend in the number of spermatogonia, spermatocytes and spermatids (Figure 3) indicated that all germinal cells are influenced by MTBE, although the effect is more prominent in the more mature cell types of spermatocytes and spermatids, and in higher dose groups. One simple explanation for this finding may be the indirect effect of decrease in the serum testosterone level in MTBE treated animals which has been reported by previous studies (Williams et al., 2000; de Peyster et al., 2003; Li et al., 2008). Furthermore, reduced number of interstitial cells in

MTBE treated groups which are testosterone secreting cells in the male reproductive system could be considered supporting evidence for the aforementioned finding. In contrast to the significant decline in spermatogenic cell numbers, no significant change was observed in the germinal epithelium height of the seminiferous tubules. This finding could be attributed to the disordered arrangements in the cells, so that lowering cell numbers was compensated by the increasing distance between cells. We observed that Sertoli cell numbers were not changed following MTBE exposure. Sertoli cells are supportive cells of the testis and their tight junctions are responsible for the formation of the blood-testis barrier. Li et al. (2007, 2009) conducted a series of *in vitro* experiments and reported that MTBE exerts toxic actions directly on spermatogenic and Sertoli cells in culture. They suggested that exposure to MTBE *in vivo* could impair spermatogenesis through toxic actions exerted directly on spermatogenic or Sertoli cells. Taken together, whether the observed changes in the number of spermatogenic cells in the present study are due to the direct effect of MTBE via passing blood-testis barrier or is an indirect effect through another mechanism such as hormonal changes should to be elucidated in the future studies. It is worth noting that a few of the previous studies on MTBE evaluated the testicular histopathology, with the majority reporting no effect (Bird et al., 1997; Williams et al., 2000; de Peyster et al., 2008; Bermudez et al., 2012). We suppose that the major reason for the observed difference in our results with the previous ones is the qualitative approach of evaluation in all previous.

About sex ratio, To the best knowledge of the authors, there is the first report about the effect of MTBE exposure of male rats on their off-spring sex ratio. We found three studies which has addressed the association of MTBE exposure with offspring sex ratio. In the first study, no treatment related change was observed in fetal sex distribution when mated CD Sprague-Dawley rats and CD-1 mice were exposed during the period of organogenesis to target concentrations of 0, 250, 1000, and 2500 ppm MTBE (Conaway et al., 1985). In the second study again with no effect on offspring sex ratio, both male and female rats were exposed by inhalation to 300, 1300 and 3400 ppm for 6 hours/day, 5 days/week before mating (12 weeks and 3 weeks, respectively) and then throughout mating and gestation (Biles et al., 1987). The third study reported that following exposure of female pregnant mice in gestational day 6-15 to various doses of MTBE by inhalation, altered sex ratio as decreasing male fetuses was observed (Bevan et al., 1997). Results of the present study showed that exposure of male rats to oral MTBE has no effect on their offspring sex ratio at birth (Table 4). Previous studies have shown that paternal exposure to gasoline (containing MTBE) in filling station workers was associated with decrease in offspring sex ratio (Ansari-Lari et al., 2004). Also, sex ratio was skewed to female births in offspring of male Sprague-Dawley rats exposed to gasoline vapor for 6 hours daily for 30 consecutive days (Ansari-Lari and Tanideh, 2009). Based on the pre-sent results, it seems that changes in sex ratio may be attributed to some gasoline components other than MTBE. However, further studies are needed to clarify this finding.

Fertility was apparently lower in the 1600 mg/kg/day MTBE compared with other experimental groups, but did not show significant difference. It is suggested that high dose of MTBE has the potential to negatively affect the fertility in male Sprague-Dawley rats and it was not significant due to limited sample size. This finding need to be investigated in studies concerning changes in sperm quantity and quality in MTBE ex-posed male rats.

In conclusion, the present study showed that paternal exposure to various doses of oral MTBE for 30 days is not associated with offspring sex ratio at birth but has the potential to negatively affect the fertility in male Sprague-Dawley rats. MTBE exposure could exert negative effect on serum testosterone, LH and GPX activity in treatment groups. A dose-related decreasing pattern was observed for the effect of MTBE exposure on sex hormones, whereas the observed effect on GPX did not show dose-dependent trend. The results of the present study, need to be clarified in the future studies may be in conjunction with investigating the genetic back ground of the host.

References

Ansari-Lari M, Saadat M, Hadi N. Influence of *GST1* null genotype on the offspring sex ratio of gasoline filling station workers. *J Epidemiol Commun Health.* (2004) 58:393-4.

Ansari-Lari M, Tanideh N. Changes in sex hormones and offspring sex ratio following gasoline exposure in male rats. *Comp Clin Path.* (2009)18:43-5.

Belpoggi F, Soffritti M, Maltoni C. Methyl-tert butyl ether (MTBE) a gasoline additive– causes testicular and lymphohaematopoietic cancers in rats. *Toxicol Ind Health.* (1995) 11:119-49.

Bermudez E, Willson, G, Parkinson, H and Dodd, D. Toxicity of methyl tertiary-butyl ether (MTBE) following exposure to Wistar rats for 13 weeks or one year 1720 via drinking water. *J. Appl. Toxicol.* (2012) 32: 687-706.

Bevan C, Tyl RW, Neeper-Bradley TL, Fisher LC, Panson R, Douglas JF, et al. Developmental toxicity evaluation of methyl tertiary-butyl ether (MTBE) by inhalation in mice and rabbits. *J Appl Toxicol.* (1997)17:S21-9.

Bird MG, Burleigh-Flayer HD, Chun JS, Douglas JF, Kneiss JJ, Andrews LS. Oncogenicity studies of in-haled methyl tertiarybutyl ether (MTBE) in CD-1 mice and F-344 rats. *J Appl Toxicol.* (1997)1 (Suppl):S45-S55.

Biles RW, Schroeder RE, Holdsworth CE. Methyl tertiary butyl ether inhalation in rats: a single generation reproduction study. *Toxicol Ind Health* (1987) 3: 519-34.

Billiti JE, Faulkner BC, Wilson BW. Absence of acute testicular toxicity of methyl-tert butyl ether and breakdown products in mice. *Bull Environ Contam Toxicol.* (2005) 75:228-35.

Clegg ED, Cook JC, Chapin RE, Foster PM, Duston GP. Leydig cells hyperplasia and adenoma formation: mechanisms and relevance to human. *Reprod Toxicol.* (1997)11:107-21.

Conaway CC, Schroeder RE, Snyder NKJ. Teratology evaluation of methyl tertiary butyl ether in rats and mice. *J Toxicol Environ Health.* (1985);16:797-809.

Day KJ, de Peyster A, Allgaier BS, Luong A, Mac-Gregor JA. Methyl t-butyl ether (MTBE) effects on the male rat reproductive endocrine axis. *Toxicologist.* (1998);42(1S):174.

de Peyster A, MacLean KJ, Stephens BA, Ahern LD, Westover CM, Rozenshteyn D. Subchronic studies in Sprague-Dawley rats to investigate mechanisms of MTBE-induced Leydig cell cancer. *Toxicol Sci.* (2003);72:31-42.

James WH. Male reproductive hazards and occupation. *Lancet* (1996);347:773.

Li, D, Liu, Q, Gong, Y, Huang, Y and Han, X. Cytotoxicity and oxidative stress study in cultured rat Sertoli cells with methyl tert-butyl ether (MTBE) exposure. *Reprod. Toxicol.,* (2009) 27: 170-176.

Li, D, Yin, H and Han, X. Methyl tert-butyl ether (MTBE)-induced cytotoxicity and oxidative stress in isolated rat spermatogenic cells. *J. Appl. Toxicol.,* (2007) 27: 10-17.

Li D, Yuan C, Gong Y, Huang Y, Han X. The effect of methyl tert-butyl ether (MTBE) on the male rat reproductive system. *Food Chem Toxicol.* (2008)46: 24-28.

Mocarelli P, Gerthoux PM, Ferrari E, Patterson DJ Jr, Kieszak SM, Brambilla P, et al. Paternal concentra-tion of dioxin and sex ratio of offspring. *Lancet* (2000) 355:1858-63.

Rosenfeld CS, Roberts RM. Maternal diet and other factors affecting offspring *sex ratio: a review. Biol Reprod.* (2004)71:1063-70.

Stern BR, Kneiss JJ. Methyl tertiary-butyl ether (MTBE): use as an oxygenate in fuels. *J Appl Toxi-col.* 1997;17:S12

USEPA. *MTBE Fact Sheet #1: Overview. EPA 510-F-97-014,* US. EPA. Washington, DC: Office of Solid Waste and Emergency Response, 1998.

Williams, TM, Cattley, RC and Borghoff, SJ. Alterations in endocrine responses in male Sprague-Dawley rats following oral administration of methyl tert-butyl ether. *Toxicol. Sci.* (2000) 54: 168-176.

Chapter 3

Lactic Acid Bacteria in the Gastrointestinal Tract: Anti-*Helicobacter pylori*

Sujitra Techo[1]
Engkarat Kingkaew[2]
and Somboon Tanasupawat[3,*]
[1]Microbial Diversity and Bioproducts Research Unit, Mahidol University, Nakhonsawan Campus, Nakhonsawan, Thailand
[2]Department of Biology, School of Science, King Mongkut's Institute of Technology Ladkrabang, Bangkok, Thailand
[3]Department of Biochemistry and Microbiology, Faculty of Pharmaceutical Sciences, Chulalongkorn University, Bangkok, Thailand

Abstract

Numerous bacterial species have been found throughout the human GI tract, which consists of the oral cavity, stomach, intestine, and colon. Lactic acid bacteria (LAB), including *Lactobacillus* (*L.*) *acidophilus*, *L. salivarius*, *L. johnsonii*, *L. crispatus*, *L. casei*, *L. paracasei*, *L. rhamnosus*, *L. reuteri*, *L. oris*, *L. vaginalis*, *L. gasseri*, *L. plantarum*, *L. buchneri*, *L. mali*, *L. ruminis*, *L. delbrueckii*, *L. sakei*, *L. fermentum*, *L. helviticus*, and *L. brevis*; *Streptococcus* (*S.*) *salivarius*, *S. pneumonia*, *S. mitis*, *S. oralis*, *S. parasanguinis*, *S. anginosus*, *S. constellatus*, *S. caprinus*, and *S. mutans*; *Leuconostoc* (*Ln.*) *mesenteroides* and *Ln. argentinum*; *Lactococcus* (*Lc.*) *lactis*; and *Enterococcus* (*En.*) strains, are distributed throughout the human GI tract. The bacterial microbiota in humans changes when infected with *Helicobacter* (*H.*) *pylori*. Moreover, antibiotics and proton pump inhibitors (PPIs) used in the regimen

---

* Corresponding Author's Email: Somboon.T@chula.ac.th.

In: Advances in Biology. Volume 6
Editor: Charles D. Grant
ISBN: 979-8-89113-712-7

influence microbial microbiota. However, alteration of bacterial diversity is rather complicated and has conflicting effects. To maintain microbial balance in the GI tract, several studies have used probiotics as adjunctive therapy to reduce the number of *H. pylori* in infected humans, replacing antibiotics alone. Probiotics have shown a higher eradication rate of *H. pylori* when compared with antibiotics alone. Bacteriocin production by LAB is one of the interesting mechanisms that can inhibit the growth of *H. pylori.* Bacteriocins and bacteriocin-like substances exhibiting antibacterial properties against *H. pylori* have been reported. These active peptides could be exploited to develop novel therapeutics for treating *H. pylori* infection.

Keywords: antibiotics, bacteriocins, lactic acid bacteria, gastrointestinal tract, gut microbiota, *helicobacter pylori*, probiotics

Abbreviation

rDNA	Ribosomal DNA;
CAG	Chronic atrophic gastritis;
GIT	Gastrointestinal Tract;
HFD	high fat diet;
MALT	Mucosa associated lymphoid tissue;
LAB	Lactic acid bacteria;
LGG	*Lactobacillus* GG;
PPI	proton pump inhibitor;
SCS	Spent culture supernatant.

1. Introduction

The human gastrointestinal (GI) microbiotas have been extensively investigated for many years and it has been found that they play a major role in human health and disease (Kolida et al., 2006; Clemente et al., 2012). The gut microbiota is considered a key "organ" that affects host biology, meaning that if the gut microbiota is compromised, the whole body is affected. Dysbiosis of the gut microbiota has been associated with the development and progression of diverse human diseases (Li et al., 2016). Lactic acid bacteria (LAB) can be found in plants or plant-derived materials, silage, fermented foods (such as yogurt, cheese, olives, pickles, and salami), as well as in the oral cavities, gastrointestinal tracts (GIT), and vaginas of humans and animals (Walter, 2008). In particular, the *Lactobacillus* species found in the GIT have received tremendous attention due to their health-promoting properties. They are commonly used as probiotics, defined by the FAO/WHO (2002) as "live microorganisms that, when administered in adequate amounts, confer a health benefit on the host."

Helicobacter pylori strains are considered etiologic factors for the development of peptic ulcer disease, gastric adenocarcinoma, and MALT lymphoma. Therapeutic schemes to eradicate them are based on double antibiotic therapy and proton pump inhibitors (PPIs). Despite many therapeutic improvements in *H. pylori* eradication treatment, it is still associated with high infection rates, especially in developed countries. Bacterial resistance and adverse events occurrence are among the most frequent causes of anti-*H. pylori* treatment failure (Homan and Orel, 2015). Around four decades ago, attempts were made to improve health status by modulating the indigenous intestinal flora using live microbial adjuncts, now called 'probiotics' (Holzapfel et al., 1998). In clinical trials, probiotics have been extensively investigated for use as adjunctive treatment for intestinal infections caused by pathogens such as *H. pylori* (Ariefdjohan, 2017). However, the application of probiotics for *H. pylori* infection therapy is supported by a low number of studies.

In this chapter, we highlighted the diversity of gut microbiota and lactic acid bacteria present in the GI tract. We reviewed important findings associated with alterations of gut microbiota in *H. pylori*-infected humans and mice, and clinical trials that use probiotics and bacteriocins as alternative or adjunctive therapies for the treatment of *H. pylori* infection.

2. Microbial Communities in Human GI Tract

The human gastrointestinal (GI) tract constitutes a bacterial ecosystem that is diverse and complex, now well recognized for its impact on human health and well-being (Doré and Corthier, 2010). The indigenous GI tract microflora has profound effects on the anatomical, physiological, and immunological development of the host (Berg, 1996). The indigenous bacteria are not randomly distributed throughout the GI tract of adults but are found at characteristic population levels in particular regions of the tract. The normal indigenous intestinal microflora consists of about 10^{14} bacteria (10 times the total number of eukaryotic cells in all tissues of the human body), representing more than 1,000 different bacterial species (Holzapfel et al., 1998; Ley et al., 2006; Qin et al., 2010). There are 200 species of indigenous microflora residing in the oral cavity. Transient bacteria (10^9 CFU/mL) are found in saliva (Berg, 1996). Intrinsic factors including pH, secretion of intestinal fluids, and transit time affect the population size and diversity of the intestinal microflora. A severe environment is produced in the stomach and duodenum. Secretion of intestinal fluids (i.e., hydrochloric acid) generates low pH (2.5 to 3.5) in the stomach and a short transit time, resulting in only low numbers of bacteria (10^1–10^3 CFU/mL of intestinal content) being detected in the human stomach and duodenum, with a majority of them being transient (Ariefdjohan et al., 2017).

Although severe conditions presented in the stomach have limited the survival of many bacteria, *Helicobacter pylori*, a spiral-shaped Gram-negative bacterium, has frequently been found in the human stomach. This bacterium has been isolated from materials of gastric sites of diverse populations throughout the world. It is considered an indigenous bacterium of the human stomach and is associated with chronic gastritis, peptic ulcers, and is likely to be an important factor in the development of gastric cancer (Owen, 1995). In addition, acid-resistant lactobacilli and streptococci could also be detected in the stomach and duodenum (Berg, 1996). The human jejunum and ileum have a higher variety and number of bacteria (10^8 CFU/mL) compared to bacteria present in the stomach and duodenum (Holzapfel et al., 1998). Lactobacilli, streptococci, coliform bacteria, bacteroides, bifidobacteria, and fusobacteria (10^4–10^8 CFU/mL) are predominant bacteria found in the small intestine. The colon is the major site for bacterial fermentation of non-digestible food components. Approximately 10^{10}-10^{12} bacteria/mL of intestinal contents reside in the colon. This microbial population includes more than 500 species belonging to more than 190 genera. The main genera dominant in the large intestine include *Bacteroides*, *Bifidobacterium* (*B.*), *Streptococcus*, *Eubacterium*, *Clostridium* (*C.*), *Fusobacterium*, *Veillonella*, *Lactobacillus*, *Proteus*, *Staphylococcus*, *Pseudomonas*, coliform bacteria, yeasts, and protozoa (10^{10}–10^{12} CFU/mL) (Ariefdjohan et al., 2017).

3. Lactic Acid Bacteria Distributed in Human GI Tract

Lactic acid bacteria (LAB), especially *Lactobacillus* strains, are widely distributed throughout the human GI tract. Traditionally, gut microbiota research relied on techniques that required the cultivation of microbes.

However, in the last two decades, culture-independent molecular approaches have been intensively applied to the study of microbial diversity in the gut ecosystem (Walter, 2008). Genome sequences for LAB have become increasingly available in recent decades, considerably expanding insight into the adaptation and functionality of these bacteria in the digestive tract. Access to genome sequence information not only facilitates molecular strategies for investigating the behavior of LAB in the intestine but also provides opportunities for high-throughput functional genomics techniques (Vaughan et al., 2005). *Lactobacillus*, *Streptococcus*, *Lactococcus*, and *Enterococcus* strains have been detected throughout the human GI tract.

To identify the autochthonous (indigenous) and allochthonous (transient) species present in the human GI tract, Reuter (2001) employed a quantitative cultivation procedure involving ten-fold dilution steps after homogenization of the samples, using a set of polyvalent media, supplemented blood agar, and various selective media. Several species of autochthonous *Lactobacillus* have been found in the mouth, including *L. acidophilus*, *L. salivarius*, *L. johnsonii*, *L. crispatus*, *L. casei*, *L. paracasei*, *L. rhamnosus*, *L. reuteri*, *L. oris*, and *L. vaginalis* (Reuter, 1965a, b; 1975). A few major species such as *L. reuteri*, *L. oris*, *L. vaginalis*, and *L. ruminis* were found to be autochthonous bacteria in the human stomach (Reuter, 1965a, b; 1975; 2001). *L. gasseri*, *L. reuteri*, *L. salivarius*, *L. casei*, *L. plantarum*, *L. buchneri*, and *L. mali* appeared to colonize the jejunum. Additionally, *Lactobacillus*, *Streptococcus*, *Lactococcus*, and *Enterococcus* strains are also present in the jejunum (Lerche and Reuter, 1962; Reuter, 1975; Hayashi et al., 2005; Wang et al., 2005). Numerous *Lactobacillus* species, *L. gasseri*, *L. reuteri*, *L. salivarius*, *L. casei*, *L. plantarum*, *L. buchneri*, and *L. mali*, exhibit much diversity in the ileum, followed by *Streptococcus* (*S. salivarius*, *S. constellatus*, *S. pneumonia, S. mitis*, and *S. oralis*) and *Enterococcus*, *Lactococcus* strains, respectively (Lerche and Reuter, 1962; Reuter, 1975; Reuter, 2001; Hayashi et al., 2005). *L. reuteri* and *L. gasseri* are the predominant species found in the human caecum and colon. Other *Lactobacillus* species, including *L. salivarius*, *L. casei*, *L. plantarum*, *L. ruminis*, *L. vaginalis*, *L. delbrueckii*, and *L. mali*, reside in the caecum. Additionally, other *Streptococcus* species, *Enterococcus* species, and *Leuconostoc mesenteroides* are also present in the caecum, colon, and recto-sigmoid colon (Lerche and Reuter, 1962; Reuter, 1975; Reuter, 2001; Heilig et al., 2002; Hayashi et al., 2005). Members of the genera *Leuconostoc*, *Pediococcus*, and *Weissella* are detected using a *Lactobacillus* group-specific PCR primer, S-G-Lab-0677-a-A-17, to selectively amplify 16S ribosomal DNA (rDNA) from lactobacilli and related lactic acid bacteria

(Heilig et al., 2002). A few *Lactobacillus* species such as *L. casei*, *L. salivarius*, and *L. rhamnosus*, are detected in infant feces, while there is a more diverse range of *Lactobacillus* species and other LAB found in adult feces, including *L. sakei, L. ruminis, L. delbrueckii, L. fermentum, L. acidophilus, L. crispatus, L. rhamnosus*, and *Ln. argentinum*. Other species such as *L. salivarius*, *L. gasseri*, *L. helveticus*, *L. paracasei*, *L. casei*, *L. curvatus*, *L. plantarum*, *L. buchneri*, *L. brevis*, and *L. reuteri* can also be found in adult feces (Štšepetova et al., 2011). However, Hayashi et al. (2002) reported that only *Streptococcus* (*S. salivarius, S. caprinus, S. anginosus, S. mutans, S. parasanguinis,* and *Streptococcus* spp.) are present in adult feces. *Lactobacillus* spp. found in elderly feces have similar diversity to adult feces. Mikelsaar et al. (2010) revealed that *L. acidophilus, L. delbrueckii* subsp. *delbrueckii, L. helviticus, L. paracasei, L. casei, L. plantarum, L. fermentum*, and *L. reuteri* have been detected in elderly feces, while *L. acidophilus, L. salivarius, L. crispatus, L. johnsonii, L. gasseri, L. delbrueckii, L. helveticus, L. ruminis* subgroup, *L. paracasei, L. casei, L. sakei, L. curvatus, L. plantarum, L. buchneri, L. brevis, L. fermentum,* and *L. reuteri* can also be found in elderly feces (Štšepetova et al., 2011). Only a few out of the >200 known *Lactobacillus* species have been consistently and repeatedly associated with the human GI tract. Recently, this number has increased to over 50 *Lactobacillus* species that were repeatedly detected in the stools of healthy volunteers. The most abundant lactobacilli include *L. casei, L. delbruckeii, L. murinus, L. plantarum, L. rhamnosus*, and *L. ruminus*. Some of these species (e.g., *L. rhamnosus* and *L. murinus*) are rarely isolated from environments outside the intestine and are considered gut-autochthonous microorganisms (Heeney et al., 2018). Large quantities of live LAB are consumed within fermented foods, but it is not yet known to what extent the LAB we ingest become members of the gut microbiome. Through the analysis of 9445 metagenomes from human samples, Pasolli et al., (2020) demonstrate that the prevalence and abundance of LAB species in stool samples are generally low and linked to age, lifestyle, and geography, with *Streptococcus thermophilus* and *Lactococcus lactis* being the most prevalent.

4. *H. pylori* Infected in Human/Mice and Changes in Gut Microbial Communities

Several factors affect the bacterial microbiota in the human stomach, including pH in the intestinal tract, oxygen tension, nutrient availability, colonic physiology, bacterial interference, etc. (Hill, 1981). One of the major factors controlling bacterial distribution in the gastrointestinal tract is the gastric acid barrier and the progression of *H. pylori*-induced chronic atrophic gastritis (CAG). The use of inhibitors of gastric acid secretion, gastrectomy, and an increase in gastric pH due to dietary indiscretion or stress reduce the acidity that usually renders the upper gastrointestinal environment unfavorable for bacterial survival (Husebye, 2005). Therefore, several studies have investigated the effect of *H. pylori* infection and proton pump inhibitor (PPI) plus antibiotic treatment on microbial diversity throughout the GI tract in both human and mouse models (Table 1). Aebischer et al. (2006) investigated changes in microbial diversity in the mouse stomach using molecular analysis. The strains of *Clostridium*, *Bacteroides/Prevotella*, *Eubacterium*, *Ruminococcus*, streptococci, lactococci, and *Escherichia coli* were increased in the stomachs of infected animals, whereas lactobacilli were decreased. However, alteration of the gastric microbiota in *H. pylori*-infected humans is still a controversial issue. Bik et al. (2006) reported that the presence of *H. pylori* did not affect the composition of the gastric community. The number of clones of *Proteobacteria, Firmicutes, Actinobacteria, Bacteroidetes*, and *Fusobacteria* phyla did not alter in humans infected with *H. pylori. H. pylori*-negative patients had a greater relative abundance of *Actinobacteria* and *Firmicutes*, whereas *H. pylori*-positive subjects had higher abundances of non-*H. pylori*, *Proteobacteria* and *Acidobacteria* (Maldonado-Contreras et al., 2011). Long-term *H. pylori* colonization affected the distribution and numbers of indigenous microflora in the stomach and duodenum. The gastric microenvironment may be unfit for lactobacilli fertility after long-term *H. pylori* infection, while enterococci, *S. aureus*, bifidobacteria, and *Bacteroides* showed colonization affected the distribution and numbers of indigenous microflora in the stomach and duodenum. After *H. pylori* infection, *Enterococcus* spp. and *Staphylococcus aureus* showed occurrences in the stomach and duodenum, while *Lactobacillus* strains showed a downtrend in the stomach. *Bacteroides* spp. significantly increased in *H. pylori*-positive gerbils (Yin et al., 2011). The 16S rRNA gene sequencing of fecal samples at 12 weeks revealed that *H. pylori*-infected mice feeding a high-fat diet (HFD) had a distinct microbiota community that clustered separately from both the HFD controls and the chow diet groups. The *H. pylori*-infected group feeding HFD had a significantly increased proportion of sequences assigned to *Firmicutes* and *Proteobacteria*, whereas reads assigned to *Bacteroidetes* and *Verrucomicrobia* were relatively reduced (He et al., 2016). The normal gastric flora in humans infected with *H. pylori* was altered upon administration of omeprazole (20 mg twice a day) and amoxicillin (1 g twice a day) for 14 days. Increases in the number of *Streptococcus milleri, S. mitior, Staphylococcus* spp., *Micrococcus*, yeasts, anaerobic Gram-positive cocci, *Actinomyces* spp., *Veillonella* spp., and *Propionibacterium* were observed at 4 weeks after treatment. *Streptococcus salivarius, S. sanguis, Lactobacillus*, and *Bifidobacterium* spp. were decreased at day 42. The eradication rate of *H. pylori* was 61% among the patients treated with 20 mg of omeprazole plus 1000 mg of amoxicillin twice a day for 14 days (Adamsson et al., 1998).

The population of bacteria belonging to *Prevotella, Atopobium* group, *Clostridium coccoides* group, *Veillonella, Lactobacillus* group, *Enterobacteriaceae*, and *Bacteroides fragilis* group showed a significant increase in acid inhibitor-treated rats compared with the non-treated rats. In contrast, no significant differences in the populations of *Bifidobacterium*, the *Eubacterium cylindroides* group, and the *C. leptum* subgroup were noted among the control and treatment groups. For humans with and without chronic atrophic gastritis (CAG), the group with CAG was found to have a larger population of all genera or groups, with the exception of Enterobacteriaceae and the *C. ramosum* group, compared with the group without CAG. Among the 10 increased genera or groups, the difference between the two groups with and without CAG was significant for *Veillonella* ($p = 0.003$) and the *Lactobacillus* group ($p = 0.002$) (Kanno et al., 2009).

A study described the changes in lower intestinal microflora induced by gastric acid reduction. In both proton pump inhibitor (PPI)-treated rat and human experiments, most genera or groups of intestinal microflora (facultative and obligate anaerobes) proliferated due to gastric acid reduction, and marked and significant increases in the Lactobacilli group and *Veillonella*, oropharyngeal bacteria, were observed. In rats, potent gastric acid inhibition led to a marked and significant increase of intestinal bacteria, including the *Bacteroides fragilis* group, while *Bifidobacterium*, a beneficial bacterial species, remained at a constant level (Kanno et al., 2009).

Before the eradication of *H. pylori*, the most abundant phyla were *Bacteroidetes* (52.09%), *Firmicutes* (32.91%), *Actinobacteria* (6.68%), and *Proteobacteria* (5.77%). At 6 months post-*H. pylori* eradication with antibiotics, the relative abundance of *Bacteroidetes* decreased to 47.82% and *Firmicutes* increased to 37.82%, as compared to baseline. Additionally, the relative abundance of both phyla *Actinobacteria* (4.86%) and *Proteobacteria* (3.69%) also reduced 6 months post-eradication. Interestingly, they also observed that the relative abundance of *Verrucomicrobia* increased markedly, from 0.07% at baseline to 3.30% 6 months post-eradication (Yap et al., 2016).

Ferreira et al. (2017) conducted a study to identify the most relevant taxa responsible for the differences between clinical diagnoses. In gastric carcinoma, an enrichment in *Proteobacteria* taxa was observed, including the genera *Phyllobacterium* and *Achromobacter*, and the families *Xanthomonadaceae* and *Enterobacteriaceae*. Although no specific genus could be identified within the *Xanthomonadaceae*, in the *Enterobacteriaceae*, the genus *Citrobacter* was identified as being significantly enriched in gastric carcinoma. Additionally, *Lactobacillus, Clostridium*, and *Rhodococcus* were also significantly more abundant in gastric carcinoma. *Helicobacter, Neisseria, Prevotella,* and *Streptococcus* were most abundant in the microbiota of patients with chronic gastritis. Moreover, at the genus level, the upper human GI tract communities in *H. pylori*-negative individuals were dominated by *Streptococcus* and *Prevotella*. Seven genera differed in abundance between communities of saliva (OAX) and stomach aspirates (SAX), with *Veillonella* being more abundant in OAX and *Eubacterium, Fusobacterium*, *Lachnoanaerobaculum*, *Olsenella*, *Parvimonas*, and *Tessarococcus* being more abundant in SAX. The higher abundance of *Firmicutes* in duodenum aspirates (DAX) can be attributed to a higher abundance of *Streptococcus*, and the lower abundance of *Bacteroidetes* and *Fusobacteria* to a lower abundance of *Prevotella* and *Fusobacterium*. Infection with *H. pylori* is typically manifested by a high abundance of this organism on the mucosa, and accordingly, stomach biopsies of individuals infected with *H. pylori* (SBH) samples revealed the presence of *Helicobacter* in amounts exceeding 50% in all infected individuals. Due to the high relative abundance of *H. pylori*, the relative abundance of all four major phyla, that is, *Actinobacteria*, *Bacteroidetes*, *Firmicutes*, and *Fusobacteria*, was low in SBH compared with stomach biopsies of individuals non-infected with *H. pylori* (SBX). Similarly, at the genus level, besides the increase in *Helicobacter*, only *Haemophilus* decreased significantly in abundance in stomach aspirates of individuals infected with *H. pylori* (SAH) samples while at the phylotype level, only Phy25 (*Haemophilus*) and Phy80 (*Campylobacter conciscus*) showed a significant decrease (Schulz et al., 2018). Overall, the gastric tissue and fluid samples from patients with gastric dysplasia or gastric cancer (GC) were predominantly colonized by members of the genera *Helicobacter* (class Epsilon-proteobacteria), *Streptococcus* (Bacilli), *Prevotella* (Bacteroidetes), *Haemophilus* (Gamma-proteobacteria), and *Neisseria* (Beta-proteobacteria) (Kwon et al., 2022).

Table 1. Changes in gastrointestinal microbiota associated with *H. pylori* infection

Model	Conditions	Changes in microbial communities			References
		Increase or higher	Decrease or lower	Unchange	
Human	*H. pylori*-induced chronic atrophic gastritis (CAG)	*Clostridium coccoides* group *Clostridium leptum* subgroup *Bifidobacterium* *Atopobium* group *Bacteroidesfragilis* group *Prevotella* *Eubacterium cylindroides* group *Fusobacterium* *Lactobacillus* group *Veillonella*	*Enterobacteriaceae* *C. ramosum* group	-	Kanno et al. (2009)
Human	*H. pylori* infection	*Proteobacteria* *Acidobacteria*	*Actinobacteria* *Firmicutes*	-	Maldonado-Contreras et al. (2011)
Human	*H. pylori* infection (6 months-post-eradication with PPI+amoxicillin 1000 mg, clarithromycin 500 mg, and pantoprazole 40 mg)	*Firmicutes* *Verrucomicrobia* *Synergistetes*	*Bacteroidetes* *Actinobacteria* *Proteobacteria* *Fusobacteria* *Euryarcheota* *Lentisphaerae* *Tenericutes*	-	Yap et al. (2016)

Model	Conditions	Changes in microbial communities			References
		Increase or higher	Decrease or lower	Unchange	
Rat	Gastric acid reduction by 100 mg/kg BW omeprazole-treatment	*Prevotella* *Atopobium* group *Clostridium coccoides* group *Veillonella* *Lactobacillus* group *Enterobacteriaceae* *Bacteroidesfragilis* group	-	*Bifidobacterium* *Eubacterium cylindroides* group *Clostridium leptum* subgroup	Kanno et al. (2009)
Gerbils	*H. pylori* infection	*Enterococcus* spp. *Staphylococcus aureus* *Bacteroides* spp.	*Lactobacillus* spp.	-	Yin et al. (2011)
Mice	*H. pylori* infection	*Clostridium coccoides* group *Clostridium* spp. *Bacteroides/Prevotella* spp. *Eubacterium* spp. *Ruminococcus* spp. *Escherichia coli* *Streptococcus* spp. *Lactococcus* spp.	*Lactobacillus* spp.	-	Aebischer et al. (2006)
Mice	*H. pylori*-infection and feeding high fat diet	*Firmicutes* *Proteobacteria*	*Bacteroidetes* *Verrucomicrobia*	-	He et al. (2016)
Human	People without *H. pylori* infection	-	-	The upper GI tract; *Streptococcus* and *Prevotella* Saliva (OAX); *Veillonella* Stomach aspirates (SAX); *Eubacterium*, *Fusobacterium*, *Lachnoanaerobaculum*, *Olsenella*, *Parvimonas* and *Tessarococcus* Duodenum aspirates (DAX); *Streptococcus*, *Prevotella* and *Fusobacterium*	Schulz et al. (2016)

Table 1. (Continued)

Model	Conditions	Changes in microbial communities			References
		Increase or higher	Decrease or lower	Unchange	
	Stomach biopsies of individuals with non-infected (SBX) or infected *H. pylori* (SBH)	*Helicobacter* (> 50%)	*Actinobacteria* *Bacteroidetes* *Firmicutes Fusobacteria*	-	
	Stomach aspirates of individuals infected with (SAH) or without *H. pylori* (SAX)	*Helicobacter*	*Haemophilus Campylobacter concicus*	-	
Human	Gastric carcinoma	genera *Phyllobacterium* and *Achromobacter* the families *Xanthomonadaceae* and *Enterobacteriaceae* *Lactobacillus, Clostridium and Rhodococcus*	-	-	Ferreira et al. (2017)
	Chronic gastritis	*Helicobacter, Neisseria, Prevotella* and *Streptococcus*	-	-	
Human	Patients with gastric dysplasia or gastric cancer (n=12)	*Helicobacter* (class *Epsilon proteobacteria*), *Streptococcus* (*Bacilli*), *Prevotella* (*Bacteroidetes*), *Haemophilus* (*Gammaproteobacteria*), and *Neisseria* (*Betaproteobacteria*)	-	-	Kwon et al. (2022)

Table 2. Antibiotics plus probiotic therapy in human studies

Subjects (No.)	Study design	Follow-Up Diagnostic Methods	Antibiotic therapy	Adjunctive therapy	Follow-Up duration	Eradication rate (%)	*P* value	Country/ References
Chlidren (83)	Randomized, Placebo-controlled, Double-blind	2 of 3 tests (^{13}C-urea breathtest, histopathology, rapid urease test)	A (25 mg/kg twice per day), C (10 mg/kg twice per day), O (0.5 mg/kg twice per day) 1 week	Antibiotics and supplemented with *Lactobacillus* GG (10^9 CFU) 1 week	4 weeks	23/34 (69%) vs 22/32 (68%)	NS	Poland/ Szajewska et al. (2009)
Children (66)	Randomized, Placebo-controlled, Double-blind	*H. pylori* stool antigen test	A (50 mg/kg/day bid), F (6 mg/kg/day bid), PPI 1 week, O (1mg/kg/day) PPI 4 weeks	Antibiotics PPI *L. acidophilus*, *L. rhamnosus*, *L. bulgaricus*, *L. casei*, *S. thermophiles*, *B. infantis*, *B. breve* (1×10^9 CFU/1 sachet) 4 weeks	8 weeks	30 (90.09%) vs 23 (69.69%)	0.04	Iran/ Ahmad et al. (2013)
Children (95)	Randomized	^{14}C-urea breath test	A (50 mg/kg/day), C (15 mg/kg/day), L (30 mg/kg/day) 2 weeks	*L. casei* 2401 *L. acidophilus* 2027 *B. lactis* 2211 (7 x 10^9 CFU/capsule) 2 weeks	8 weeks	20(68.9%) vs 18(66.6%)	0.78	Turkey/Akcam et al. (2015)
Adult (160)	Randomized	Histology and CLO test or ^{13}C-urea breath test	L (60 mg/day) A (2 mg/day) C (500 mg/day) 1 week	Antibiotics 1 week followed by *Lactobacillus* and *Bifidobacterium* species (AB-Yogurt, 5 x 10^9CFU/bottle) 4 weeks	8 weeks	^{ITT}91% vs. 78%	0.05	Sheu et al. (2002)
Adult (70)	Randomized, Placebo-controlled	13C-urea breathtest	-	AB-yogurt containing *L. acidophilus* La5, *B. lactis* Bb12, *L. bulgaricus* and *S. thermophilus* (10^7CFU/mL) 6 weeks	8 weeks	36.2% vs 28.2%	0.05	Wang et al. (2004)

Table 2. (Continued)

Subjects (No.)	Study design	Follow-Up Diagnostic Methods	Antibiotic therapy	Adjunctive therapy	Follow-Up duration	Eradication rate (%)	*P* value	Country/ References
Adult (229)	Randomized	^{13}C-urea breath and stool antigen test	R (10 mg/kg/bid) A (750 mg/kg/bid) C (200 mg/kg/bid) 1 week	Antibiotics *L. gasseri* OLL2716 (>10^9 CFU) 4 weeks	8 weeks	^{ITT}82.6% vs69.3% ^{PP}85.6%vs74.5%	0.018 0.041	Japan/ Deguchi et al. (2011)
Adult (234)	Randomized	^{13}C or ^{14}C urease breath test	O (20 mg bid) C (500 mg bid) A (1000 mg bid) 1 week (OCA group)	*L. acidophilus* two week (3 x 10^7 CFU) followed by antibiotics one week (POCA group) and antibiotics one week followed by *L. acidophilus* two week (3 x 10^9 CFU) (OCAP group)	4 weeks	^{ITT}POCA (62/78; 79.5%), OCAP (61/77; 79.2%) and OCA group (48/79; 60.8%) ^{PP}POCA (62/76; 81.6%) OCAP (61/74; 82.4%), and OCA (48/78; 61.5%)	0.015 0.014	China/ Du et al. (2012)
Adult (338)	Randomized, Placebo-controlled, Double-blind	^{13}C-urea breath test and enzyme immunoassay serology	L (30 mg b.d.) C (500 mg b.d.) A (1 g b.d.) 1 week	Antibiotics 1 week, LGG (ATCC 53103), *L. rhamnosus* LC (DSM7061) (6 x 10^8 CFU/mL), *P. freudenreichii* ssp. *shermanii* JS (DSM 7076) 7 x 10^8 CFU/mL) and *B. breve* Bb99 (DSM 13692) 7 x 10^6 CFU/mL) 3 weeks	6 weeks	21/23 (91%) vs 19/24 (79%)	0.42	Myllyluoma et al. (2005)

Subjects (No.)	Study design	Follow-Up Diagnostic Methods	Antibiotic therapy	Adjunctive therapy	Follow-Up duration	Eradication rate (%)	*P* value	Country/ References
Adult (40)	Placebo-controlled Double-blind	Endoscopy	R (20 mg, twicea day) A (1 g, twice daily) for 5 days followedby R (20 mg twice a day) plus C (500 mg, twice daily), and T (500 mg, twice daily) for the next 5 days Sequential treatment 10 days	*L. reuteri* ATCC 55730 (10^8 CFU/tablet) 4 weeks, Sequential treatment 10 days	4 weeks	88% vs. 82%	0.8	Francavilla et al. (2008)
Adult (468)	-	-	Standard primary eradication therapy for *H. pylori*; proton pump inhibitors, amoxicillin, or clarithromycin (PPI group), vonoprazan, amoxicillin, or clarithromycin (VPZ group), and proton pump inhibitors, amoxicillin, or clarithromycin/probiotics (Miya-BM®) (PPI + MBM group)	Miya-BM® (MBM) (*Clostridium butyricum* MIYAIRI 588)	-	87.1% vs. 70.1%	0.0008	Mukai et al. (2020)
Adult (664)	Placebo-controlled study	either histological assessment of gastric biopsies or ^{13}C-Urea Breath Test	10-day proton pump inhibitor containing non-bismuth quadruple therapeutic regimen for *H. pylori* eradication (omeprazole 20 mg, amoxycillin 1 g, clarithromycin 500 mg, and metronidazole 500 mg all twice daily orally)	Four probiotic strains; *L. acidophilus*, *L. plantarum*, *B. lactis*, and *Saccharomyces boulardii*	6 weeks	92.0% vs. 86.8%	0.028	Viazis et al. (2022)

Table 2. (Continued)

Subjects (No.)	Study design	Follow-Up Diagnostic Methods	Antibiotic therapy	Adjunctive therapy	Follow-Up duration	Eradication rate (%)	*P* value	Country/ References
Adult (270)	Randomized, double-blind placebo-controlled trial	^{13}C-urea breath test	Standard triple regimen for *H. pylori* eradication	*E. faecium* 4.5×10^8 and *Bacillus subtilis* 5.0×10^7 (thrice daily) for 28 days	6 months	77.1% vs 72.4%	0.48	Lim et al. (2023)

A, amoxicillin;
C, clarithromycin;
F, furazolidone;
L, lansoprazole;
O, omeprazole;
R, rabeprazole;
T, tinidazole;
NS, not significant;
^{ITT}, intention-to-treat analysis;
^{PP}, Per-protocol analysis.

5. Probiotics Effected on *H. pylori*: Clinical Trials

H. pylori is the major cause of chronic gastritis and a risk factor for peptic ulcer and gastric cancer (Myllyluoma et al., 2005). Currently, available experimental and clinical studies indicate that probiotics show promise for future applications in the management of gastric ulcers (Khoder et al., 2016). Probiotics are live microorganisms that, when administered in adequate amounts, confer health benefits on the host (FAO/WHO, 2002). They have a positive effect on the eradication of *H. pylori* infection. Adjuvant therapy with probiotics is recommended to reduce the frequency of antibiotic-induced side effects during treatment with antibiotics. The clinical trials investigating the effect of probiotics on *H. pylori* eradication rates as an adjuvant to eradicating regimens are summarized in Table 2.

A double-blind, placebo-controlled, randomized trial was conducted to compare a 7-day triple eradication regimen consisting of two antibiotics (amoxicillin tablets, 25 mg/kg twice per day, and clarithromycin tablets, 10 mg/kg twice per day) plus a proton pump inhibitor (omeprazole capsules, 0.5 mg/kg twice per day) supplemented with *Lactobacillus rhamnosus* GG (LGG) (10^9 CFU) or placebo in 83 children with *H. pylori* infection confirmed by 2 of 3 tests (^{13}C-urea breath test, histopathology, rapid urease test). The groups did not differ with respect to *H. pylori* eradication rates. Of the 34 children in the LGG group, 23 (69%) experienced eradication, compared with 22 of 32 children (68%) in the placebo group (Szajewska et al., 2009). Ahmad et al. (2013) conducted a double-blind, randomized, placebo-controlled study to investigate the effect of probiotics on *H. pylori* eradication in children. *H. pylori*-positive children were treated with triple therapy (omeprazole + amoxicillin + furazolidone) and randomly allocated to receive either probiotic or placebo. The rate of *H. pylori* eradication was significantly higher in the probiotic group (90.09%) compared to the placebo group (69.69%). Ninety-five children with dyspepsia were screened for *H. pylori* infection, with 61 out of 95 (64.2%) found positive. The patients were divided into two groups: group one (n=31) received standard triple therapy (lansoprazole, amoxicillin, and clarithromycin for 2 weeks), and group two (n=30) received standard triple therapy plus probiotics (*L. casei, L. acidophilus*, and *B. lactis*). There was no significant difference between group one (68.9%) and group two (66.6%) (P=0.78) (Akcam et al., 2015). One hundred and sixty *H. pylori*-infected patients were randomized into a triple-plus-yogurt group or a triple-only group, receiving 1 week of triple therapy with and without supplements of AB-Yogurt, respectively. In the triple-plus-yogurt group, AB-Yogurt was continued for 4 weeks after triple therapy. Eight weeks later, patients were assessed for the success of *H. pylori* eradication. By intention-to-treat analysis, the triple-plus-yogurt group (91%) had a higher *H. pylori* eradication rate than the triple-only group (78%) ($P < 0.05$) (Sheu et al., 2002).

Forty-seven of 338 volunteers were *H. pylori*-infected patients. In the antibiotic-treated group, patients received amoxicillin, clarithromycin, and lansoprazole for 7 days. The antibiotic plus probiotic-treated group received the same antibiotics for 1 week and probiotic therapy (*L. rhamnosus* GG, *L. rhamnosus* LC705, *B. breve* Bb99, and *P. freudenreichii* subsp. *shermanii* JS) or a placebo during *H. pylori* eradication for 3 weeks. The *H. pylori* eradication rate was non-significantly higher in the group receiving probiotic therapy (91% vs. 79%, $P = 0.42$) (Myllyluoma et al., 2005). In a double-blind, placebo-controlled study, 40 *H. pylori*-positive subjects were given *L. reuteri* once a day for 4 weeks or placebo and followed by sequential treatment for 10 days. After the probiotic/placebo supplementation, none of the patients had eradicated the *H. pylori* infection. After the 10-day sequential regimen, there was no difference in the rate of *H. pylori* eradication in those who had previously received probiotic or placebo (88% vs. 82%; $P = 0.8$) (Francavilla et al., 2008). A total of 229 patients were randomized into either a 1-week triple therapy of rabeprazole (10 mg bid), amoxicillin (750 mg bid), and clarithromycin (200 mg bid) or triple therapy plus *L. gasseri*-containing yogurt. In the yogurt-plus-triple therapy groups, yogurt containing *L. gasseri* OLL2716 (112 g) was given twice daily for 4 weeks (3 weeks pretreatment and also 1 week during eradication therapy). After 8 weeks, overall eradication (intention to treat/per protocol) was 69.3/74.5% for the triple-only group and 82.6/85.6% for the yogurt-plus-triple group ($P = 0.018/P = 0.041$). This study suggested that a 4-week treatment with *L. gasseri*-containing yogurt improves the efficacy of triple therapy in patients with *H. pylori* infection (Deguchi et al., 2011). Two hundred and thirty-four *H. pylori* positive gastritis patients were randomized to one-week standard triple therapy (omeprazole 20 mg bid, clarithromycin 500 mg bid, and amoxicillin 1000 mg bid; OCA group, n = 79); two weeks of pre-treatment with probiotics, containing 3×10^7 *L. acidophilus* per day, prior to one week of triple therapy (POCA group, n = 78); or one week of triple therapy followed by two weeks of the same probiotics (OCAP group, n = 77). PP analysis involved 228 patients, 78 in the OCA, 76 in the POCA, and 74 in the OCAP group. Successful eradication was observed in 171 patients; by PP analysis, the eradication rates were significantly higher (P = 0.007 each) in the POCA (62/76; 81.6%) and OCAP (61/74; 82.4%) groups than in the OCA group (48/78; 61.5%). ITT analysis also showed that eradication rates were significantly higher in the POCA (62/78; 79.5%) and OCAP (61/77; 79.2%) groups than in the OCA group (48/79; 60.8%) (P = 0.014 and P = 0.015) (Du et al., 2012). Interestingly, regular intake of yogurt containing *B. lactis* Bb12 and *L. acidophilus* La5 without antibiotic treatment effectively suppressed *H. pylori* infection in humans. Fifty-nine adult volunteers infected with *H. pylori* were given AB-yogurt (10^7 colony-forming units of both La5 and Bb12 per mL) twice daily after a meal for 6 weeks. Eleven subjects positive for *H. pylori* infection were treated with milk placebo as control subjects. After 8 weeks of therapy, ^{13}C-UBT was decreased from 36.2% to 28.2% ($P < 0.05$) (Wang et al., 2004). Patients with *H. pylori* infection who received standard treatment were divided into three groups based on their treatment regime, either proton pump inhibitors, amoxicillin, or clarithromycin (PPI group), vonoprazan, amoxicillin, or clarithromycin (VPZ group), and proton pump inhibitors, amoxicillin, or clarithromycin/probiotics (Miya-BM®) (PPI + MBM group). The eradication rate of *H. pylori* was significantly higher in the PPI + MBM group (87.1%) than in the PPI group (70.1%) (Mukai et al., 2020). All patients in participating hospitals received a 10-day proton pump inhibitor containing non-bismuth quadruple therapeutic regimen for *H. pylori* eradication (omeprazole 20 mg, amoxicillin 1 g, clarithromycin 500 mg, and metronidazole 500 mg all twice daily orally) and were randomized to receive either probiotics (group A) or placebo (group B). The probiotic used combined four probiotic strains, i.e., *L. acidophilus*, *L. plantarum*, *B. lactis*, and *Saccharomyces boulardii*. *H. pylori* was successfully eradicated in 303 patients in group A (92.0%) and 291 patients in group B (86.8%) (P = 0.028) (Viazis et al., 2022). Two hundred seventy patients who received a standard triple regimen for *H. pylori* eradication were further randomized to receive a probiotic as adjunctive therapy (*E. faecium* 4.5×10^8 and *Bacillus subtilis* 5.0×10^7; Medilac-S®, Hanmi Pharmaceuticals, Seoul, Korea) or a placebo (one tablet thrice daily) for 28 days following *H. pylori* eradication. Eradication rates in the probiotic and placebo groups were 77.1% and 72.4%, respectively (P = 0.48) (Lim et al., 2023).

6. Bacteriocins, Active Peptides against *H. pylori*

Recently, there has been a growing interest in exploring bacteriocins and their potential in human and veterinary medicine (Ghobadi et al., 2021). Bacteriocins are ribosomally-synthesized peptides or proteins with antimicrobial activity produced by various groups of bacteria, including many members of LAB (Zendo, 2013). They are extracellularly secreted bioactive peptides that exert a bactericidal or bacteriostatic effect on other bacteria. These small, heat-stable peptides can have a narrow or broad target spectrum, and many are active in the nanomolar range (Molloy et al., 2002). Many bacteriocins from LAB exhibit antagonistic activity against Gram-positive bacteria such as *Staphylococcus aureus*, *En. faecalis*, *Listeria monocytogenes*, *Bacillus cereus*, and *Bacillus coagulans* (Lim and Im, 2009; Lee et al., 2007; Huang et al., 2009; Masuda et al., 2011; Woraprayote et al., 2023). Only a few bacteriocins released from LAB (e.g., *En. faecium*, *Lc. lactis*, *En. hirae*, and *En. canis*) possess inhibition against Gram-negative bacteria, including various *Escherichia coli* strains (Ponce et al., 2008). Bacteriocins 7293A and 7293B produced by *Weissella hellenica* BCC 7239 have been shown to inhibit *Pseudomonas aeruginosa*, *Aeromonas hydrophila*, *Salmonella* Typhimurium, and *E. coli* (Woraprayote et al., 2015). Although studies on the antimicrobial activity of bacteriocins produced by LAB have increased continuously for almost a hundred years, investigations into the effects of these substances on anti-*H. pylori* activity and their characteristics are limited. *L. plantarum* BK10, *L. brevis* BK11, *L. acidophilus* BK13, *L. paracasei* BK57, and *En. faecalis* BK61 strains showed bactericidal activity against *H. pylori*, significantly correlated with bacteriocin production (Lim, 2014). The crude bacteriocins of *L. brevis* BK11 and *En. faecalis* BK61 strains dramatically decreased the viability of *H. pylori*. Consequently, these two strains were chosen for further bacteriocin purification and characterization. Lim (2015) investigated the antimicrobial effects of purified bacteriocins (BK11, 6.5 kDa and BK 61, 4.5 kDa) isolated from *L. brevis* BK11 and *En. faecalis* BK61 on *H. pylori*.

Bulgaricin BB18, with a molecular mass of 4.2 kDa, produced by *L. delbrueckii* subsp. *bulgaricus* BB18, strongly inhibited *H. pylori* (Simova et al., 2009). Nisin A; lacticins A164, BH5, JW3, and NK24; pediocin PO2; and leucocin K were selected to test their antibacterial activity against various *H. pylori* strains. Among these bacteriocins, lacticins A164 and BH5 produced by *Lc. lactis* subsp. *lactis* A164 and *Lc. lactis* BH5, respectively, exhibited the strongest antibacterial activity against *H. pylori* strains. Pediocin PO2 was less active than the lacticins against four strains of *H. pylori*, and leucocin K showed no inhibition toward *H. pylori* ATCC 43504 (Kim et al., 2003). The spent culture supernatant (SCS) of a selected strain *En. faecium* TM39 significantly inhibited the viability and urease activity of *H. pylori in vitro*. This antagonistic effect, in addition to pH and lactic acid, suggests that bacteriocin-like substances in TM39-SCS might play a major role (Tsai et al., 2004). The proteinaceous compounds produced from *L. fermentum* P43-01 showed the broad spectrum of antimicrobial activity against *H. pylori* strains MS83 and BK364 which isolated from patients suffering from peptic ulcer diseases, and gastric cancer (Techo et al., 2019). Furthermore, antimicrobial peptides secreted from six strains of *Bifidobacterium* species against *H. pylori* have been reported. These antagonistic effects were related to heat-stable proteinaceous compounds, resistant to heating at 100°C for 10 minutes but sensitive to proteases (Collado et al., 2005). *L. pentosus* SLC13 exhibited an inhibitory effect against *H. pylori* J99 (Thuy et al., 2022), and *L. crispatus* FSCDJY67L3 showed eradication potential (Hong et al., 2023). A novel *Lactococcus* species strain NH2-7C, which produced nisin A, exhibited anti-*H. pylori* activity (Kingkaew et al., 2023). Chemical synthesis of plantaricin NC8 (PLNC8), a bacteriocin secreted from *L. plantarum* ZJ316, demonstrated significant anti-*H. pylori* ZJC03 potential. It resulted in a notable reduction in urease activity, with MIC of 80 μM, and inhibited the growth of *H. pylori* ZJC03 by disrupting its structure (Liang et al., 2022). These findings suggest that bacteriocins could be harnessed to develop novel therapeutics aimed at reducing or replacing the use of antibiotics in the treatment of *H. pylori* infection.

Conclusion

In conclusion, the human GI microbiota, particularly the LAB, plays a significant role in human health and disease. Alterations in the gut microbiota, known as dysbiosis, have been linked to various human diseases, including those caused by *H. pylori* infection. *H. pylori* infection, a major cause of chronic gastritis, peptic ulcer disease, and gastric cancer, can disrupt the balance of microbial communities in the gastrointestinal tract. Studies have shown that the composition of the gut microbiota is affected by *H. pylori* infection and its treatment, including antibiotic therapy and PPIs. While some bacteria increase in abundance, others decrease, leading to shifts in microbial diversity throughout the GI tract. Furthermore, the use of probiotics as adjunctive therapy for *H. pylori* infection has shown promise in clinical trials, with some studies reporting improved eradication rates and reduced antibiotic-induced side effects. Overall, understanding the complex interactions between *H. pylori*, the gut microbiota, and probiotics is crucial for developing effective strategies for the prevention and treatment of *H. pylori*-related diseases. Further research is needed to elucidate the mechanisms underlying these interactions and optimize probiotic interventions for *H. pylori* infection therapy.

References

Adamsson, I., Charlotta Edlund, Rein Seensalu, Svante Sjostedt, and Carl Erik Nord. "The normal gastric microflora and *Helicobacter pylori*; before, during and after treatment with omeprazole and amoxicillin." *Clinical Microbiology and Infection* 4(6), (1998): 308–315. https://doi.org/10.1111/j.1469-0691.1998.tb00064.x.

Aebischer, T., André Fischer, Anna Walduck, Cord Schlötelburg, Mirko Lindig, Sören Schreiber, Thomas F. Meyer, Stefan Bereswill, and Ulf B. Göbel. "Vaccination prevents *Helicobacter pylori*-induced alterations of the gastric flora in mice." *FEMS Immunology and Medical Microbiology* 46, (2006): 221–229. https://doi.org/10.1111/j.1574-695X.2005.00024.x.

Ahmad, K., Fatemeh Farahmand, Mehri Najafi, and Maryam Shoaran. "Probiotics for the treatment of pediatric *Helicobacter pylori* infection: a randomized double blind clinical trial." *Iranian Journal of Pediatrics* 23(1), (2013): 79–84.

Akcam, M., Tugba Koca, Hakan Salman, and Nermin Karahan. "The effects of probiotics on treatment of *Helicobacter pylori* eradication in children." *Saudi Medical Journal* 36(3), (2015): 286–290. https://doi.org/10.15537/smj.2015.3.10124.

Ariefdjohan, M. W., Abby Dilk, Onikia N. Brown-Esters, and Dennis A. Savaiano. "Intestinal microflora and diet in health." In *Nutrition in the Prevention and Treatment of Disease, 3rd*, edited by Coulston, A. M., Boushey, C. J., Ferruzzi, M. G., 811-33. Elsevier Inc., 2017. http://dx.doi.org/10.1016/B978-0-12-802928-2.00037-0.

Berg, R. D. "The indigenous gastrointestinal microflora." *Trends in Microbiology* 4(11), (1996): 430–435. https://doi.org/10.1016/0966-842X(96)10057-3.

Bik, E. M., Paul B. Eckburg, Steven R. Gill, Karen E. Nelson, Elizabeth A. Purdom, Fritz Francois, Guillermo Perez-Perez, Martin J. Blaser, and David A. Relman. "Molecular analysis of the bacterial microbiota in the human stomach." *Proceedings of the National Academy of Sciences of the United States of America* 103(3), 2006: 732–737. https://doi.org/10.1073/pnas.0506655103.

Clemente, J. C., Luke K. Ursell, Laura Wegener Parfrey, and Rob Knight. "The impact of the gut microbiota on human health: an integrative view." *Cell* 148(6), (2012): 1258–1270. https://doi.org/10.1016/j.cell.2012.01.035.

Collado, M. C., Ana I. González, Rosa Sanz González, Manuel A. C. Rodríguez Hernández, María Antonia Ferrús, and Yolanda Sanz. "Antimicrobial peptides are among the antagonistic metabolites produced by *Bifidobacterium* against *Helicobacter pylori*." *International journal of antimicrobial agents* 25, (2005): 385–391. https://doi.org/10.1016/j.ijantimicag.2005.01.017.

Deguchi, R., Hidemasa Nakaminami, Emiko Rimbara, Norihisa Noguchi, Masanori Sasatsu, Takayoshi Suzuki, Masashi Matsushima, Jun Koike, Muneki Igarashi, Hideki Ozawa, Ryuki Fukuda, and Atsushi Takagi. "Effect of pretreatment with *Lactobacillus gasseri* OLL2716 on first-line *Helicobacter pylori* eradication therapy." *Journal of*

Gastroenterology and Hepatology 27, (2012): 888–892. https://doi.org/10.1111/j.1440-1746.2011.06985.x.

Doré, J. and Gerard Corthier. "The human intestinal microbiota." *Gastroentérologie Clinique et Biologique* 34(Suppl. 1), (2010): S7–S15. https://doi.org/10.1016/S0399-8320(10)
70015-4.

Du, Y. Q., Tun Su, Jian-Gao Fan, Yu-Xia Lu, Ping Zheng, Xing-Hua Li, Chuan-Yong Guo, Ping Xu, Yan-Fang Gong, and Zhao-Shen Li. "Adjuvant probiotics improve the eradication effect of triple therapy for *Helicobacter pylori* infection." *World Journal of Gastroenterology* 18(43), (2012): 6302–6307. https://doi.org/10.3748/wjg.v18.i43.6302.

FAO/WHO. *Guidelines for the evaluation of probiotics in food.* London, Ontario: Food and Agriculture Organization of the United Nations and World Health Organization Working Group Report, 2002. 1–11.

Ferreira, R. M., Joana Pereira-Marques, Ines Pinto-Ribeiro, Jose L. Costa, Fatima Carneiro, Jose C. Machado, Ceu Figueiredo. "Gastric microbial community profiling reveals a dysbiotic cancer-associated microbiota." *Gut* 67(2), 2018: 226–236. https://doi.org/10.1136/gutjnl-2017-314205.

Francavilla, R., Elena Lionetti, Stefania Paola Castellaneta, Anna Maria Magistà, Giovanni Maurogiovanni, Nunzia Bucci, Angela De Canio, Flavia Indrio, Luciano Cavallo, Enzo Ierardi, and Vito Leonardo Miniello. "Inhibition of *Helicobacter pylori* infection in humans by *Lactobacillus reuteri* ATCC 55730 and effect on eradication therapy: a pilot study." *Helicobacter* 13, (2008): 127–134. https://doi.org/10.1111/j.1523-5378.2008.00593.x.

Ghobadi, E., Zahra Ghanbarimasir, and Saeed Emami. "A review on the structures and biological activities of anti-*Helicobacter pylori* agents." *European Journal of Medicinal Chemistry.* 223, 113669 (2021). https://doi.org/10.1016/j.ejmech.2021.113669.

Hayashi, H., Rei Takahashi, Takahiro Nishi, Mitsuo Sakamoto, and Yoshimi Benno. "Molecular analysis of jejunal, ileal, caecal and rectosigmoidal human colonic microbiota using 16S rRNA gene libraries and terminal restriction fragment length polymorphism." *Journal of Medical Microbiology* 54, (2005): 1093–1101. https://doi.org/10.1099/jmm.0.45935-0.

He, C., Zhen Yang, Dandan Cheng, Chuan Xie, Yin Zhu, Zhongming Ge, Zhijun Luo, and Nonghua Lu. "*Helicobacter pylori* infection aggravates diet-induced insulin resistance in association with gut microbiota of mice." *EBioMedicine* 12, (2016): 247–254. https://doi.org/10.1016/j.ebiom.2016.09.010.

Heeney, D. D., Mélanie G. Gareau, and Maria L. Marco. "Intestinal *Lactobacillus* in health and disease, a driver or just along for the ride?" *Current Opinion in Biotechnology* 49, (2018): 140–147. https://doi.org/10.1016/j.copbio.2017.08.004.

Heilig, H. G. H. J., Erwin G. Zoetendal, Elaine E. Vaughan, Philippe Marteau, Antoon D. L. Akkermans, and Willem M. de Vos. "Molecular Diversity of *Lactobacillus* spp. and other lactic acid bacteria in the human intestine as determined by specific amplification of 16S ribosomal DNA." *Applied and Environmental Microbiology* 68(1), (2002): 114–123. https://doi.org/10.1128/AEM.68.1.114-123.2002.

Hill, M. J. "Diet and the human intestinal bacterial flora." *Cancer research* 41, (1981): 3778–3780.

Holzapfel, W. H., Petra Haberer, Johannes Snel, Ulrich Schillinger, and Jos H. J. Huis in't Veld. "Overview of gut flora and probiotics." *International journal of food microbiology* 41, (1998): 85–101. https://doi.org/10.1016/S0168-1605(98)00044-0.

Homan, M. and Rok Orel. "Are probiotics useful in *Helicobacter pylori* eradication?" *World Journal of Gastroenterology* 21(37), (2015): 10644–10653. https://doi.org/10.3748/wjg.v21.i37.10644.

Hong, Q., Jidong Wang, Huayue Zhang, Xiaoming Liu, and Zhenmin Liu. "Study of the effect of *Lactobacillus crispatus* FSCDJY67L3 on *Helicobacter pylori* eradication: a double-blind randomized controlled clinical trial." *Frontiers in Immunology* 14, (2023): 1265995.

Huang, Y., Yunbo Luo, Zhengyuan Zhai, Hongxing Zhang, Chaoxiang Yang, Hongtao Tian, Zheng Li, Jiannan Feng, Hui Liu, Yanling Hao. "Characterization and application of an *anti-Listeria* bacteriocin produced by *Pediococcus pentosaceus* 05-10 isolated from Sichuan Pickle, a traditionally fermented vegetable product from China." *Food Control* 20, (2009): 1030–1035. https://doi.org/10.1016/j.foodcont.2008.12.008.

Husebye, E. "The pathogenesis of gastrointestinal bacterial overgrowth." *Chemotherapy* 51(Suppl. 1), (2005): 1–22. https://doi.org/10.1159/000081988.

Kanno, T., Takahiro Matsuki, Masashi Oka, Hirotoshi Utsunomiya, Kenichi Inada, Hirohito Magari, Izumi Inoue, Takao Maekita, Kazuki Ueda, Shotaro Enomoto, Mikitaka Iguchi, Kimihiko Yanaoka, Hideyuki Tamai, Shigeru Akimoto, Koji Nomoto, Ryuichiro Tanaka, and Masao Ichinose. "Gastric acid reduction leads to an alteration in lower intestinal microflora." *Biochemical and Biophysical Research Communications* 381, (2009): 666–670. https://doi.org/10.1016/j.bbrc.2009.02.109.

Khoder, G., Asma A. Al-Menhali, Farah Al-Yassir, and Sherif M. Karam. "Potential role of probiotics in the management of gastric ulcer." *Experimental and Therapeutic Medicine* 12, (2016): 3–17. https://doi.org/10.3892/etm.2016.3293.

Kim, T. S., Ji-Woon Hur, Myeong-Ae Yu, Chan-Ick Cheigh, Kyung-Nam Kim, Jae-Kwan Hwang, and Yu-Ryang Pyun. "Antagonism of *Helicobacter pylori* by bacteriocins of lactic acid bacteria." *Journal of Food Protection* 66, (2003): 3–12. https://doi.org/10.4315/0362-028x-66.1.3.

Kingkaew, E., Weerapong Woraprayote, Auttaporn Booncharoen, Kanidta Niwasabutra, Thitiphorn Janyaphisan, Ratha-Korn Vilaichone, Yoshio Yamaoka, Wonnop Visessanguan and Somboon Tanasupawat. "Functional genome analysis and anti-*Helicobacter pylori* activity of a novel bacteriocinogenic *Lactococcus* sp. NH2-7C from Thai fermented pork (Nham)." *Scientific Reports* 13, (2023): 20362. https://doi.org/10.3389/fcimb.2023.1120789.

Kolida, S., Delphine M. Saulnier, and Glenn R. Gibson. "Gastrointestinal microflora: probiotics." *Advances in Applied Microbiology* 59, (2006): 187–219. https://doi.org/10.1016/S0065-2164(06)59007-0.

Kwon, S. K., Jun Chul Park, Kwang H. Kim, Jaekyung Yoon, Yejin Cho, Buhyun Lee, Jin-Jae Lee, Haengdueng Jeong, Yeseul Oh, Sung-Hee Kim, So Dam Lee, Bo Ram

Hwang, Yusook Chung, Jihyun F. Kim, Ki Taek Nam, Yong Chan Lee. "Human gastric microbiota transplantation recapitulates premalignant lesions in germ-free mice." *Gut* 71(7), (2022): 1266–1276. https://doi.org/10.1136/gutjnl-2021-324489.

Lee, K. H., Jae-Yong Park, Seon-Ju Jeong, Gun-Hee Kwon, Hyong-Joo Lee, Hae Choon Chang, Dae Kyun Chung, Jong-Hoon Lee, and Jeong Hwan Kim. "Characterization of paraplantaricin C7, a novel bacteriocin produced by *Lactobacillus paraplantarum* C7 isolated from Kimchi." *Journal of Microbiology and Biotechnology* 17(2), (2007): 287–296.

Lerche, M., and G. Reuter. "Das Vorkommen aerob wachsender gram-positiver Stäbchen des Genus *Lactobacillus* Beijerinck im Darminhalt erwachsener Menschen. (Gleichzeitig ein Beitrag zur genaueren Kenntnis der aeroben Laktobazillen)." *Zentralblatt für Bakteriologie, Parsitenkunde, Infektionskrankheiten und Hygiene, Abteilung 1. Originale*, 185, (1962): 446–481.

Ley, R. E., Daniel A. Peterson, and Jeffrey I. Gordon. "Ecological and evolutionary forces shaping microbial diversity in the human intestine." *Cell* 124, (2006): 837–848. https://doi.org/10.1016/j.cell.2006.02.017.

Li, D., Pan Wang, Pengpu Wang, Xiaosong Hu, and Fang Chen. "The gut microbiota: A treasure for human health." *Biotechnology advances* 34, (2016): 1210–1224. https://doi.org/10.1016/j.biotechadv.2016.08.003

Liang, Y., Jiaqian Yan, Ziqi Chen, Qing Gu, and Ping Li. "Antibacterial effects of Bacteriocin PLNC8 against *Helicobacter pylori* and its potential mechanism of action." *Foods* 11, (2022): 1235. https://doi.org/10.3390/foods11091235.

Lim, E. S. "Purification and characterization of two bacteriocins from *Lactobacillus brevis* BK11 and *Enterococcus faecalis* BK61 showing anti-*Helicobacter pylori* activity." *Journal of the Korean Society for Applied Biological Chemistry* 58(5), (2015): 703–714. https://doi.org/10.1007/s13765-015-0094-y.

Lim, N. R., Soo Yeon Choi, and Woo Chul Chung. "Probiotic supplementation for treatment of *Helicobacter pylori* infection: a double-blind randomized clinical trial." *The Korean Journal of Helicobacter and Upper Gastrointestinal* Research 23(1), (2023): 34–41. https://doi.org/10.7704/kjhugr.2022.0051.

Lim, S. M. and Dong-Soon Im. "Screening and characterization of probiotic lactic acid bacteria isolated from Korean fermented foods." *Journal of Microbiology and Biotechnology* 19(2), (2009): 178–186. https://doi.org/10.4014/jmb.0804.269.

Lim, S. M. "Anti-*Helicobacter pylori* activity of antimicrobial substances produced by lactic acid bacteria isolated from Baikkimchi." *Journal of the Korean Society for Applied Biological Chemistry* 57(5), (2014): 621−630. https://doi.org/10.1007/s13765-014-4198-6.

Maldonado-Contreras, A., Kate C. Goldfarb, Filipa Godoy-Vitorino, Ulas Karaoz, Mónica Contreras, Martin J. Blaser, Eoin L. Brodie, and Maria G. Dominguez-Bello. "Structure of the human gastric bacterial community in relation to *Helicobacter pylori* status." *International Society for Microbial Ecology* 5, (2011): 574–579. https://doi.org/10.1038/ismej.2010.149.

Masuda, Y., Hiroshi Ono, Hiroshi Kitagawa, Haruo Ito, Fuqin Mu, Naruhiko Sawa, Takeshi Zendo, and Kenji Sonomoto. "Identification and characterization of leucocyclicin Q, a novel cyclic bacteriocin produced by *Leuconostoc mesenteroides* TK41401."

Applied and Environmental Microbiology 77(22), (2011): 8164–8170. https://doi.org/10.1128/AEM.06348-11.

Mikelsaar, M., Jelena Štšepetova, Pirje Hütt, Helgi Kolk, Epp Sepp, Krista Lõivukene, Kersti Zilmer, and Mihkel Zilmer. "Intestinal *Lactobacillus* sp. is associated with some cellular and metabolic characteristics of blood in elderly people." *Anaerobe* 16, (2010): 240–246. https://doi.org/10.1016/j.anaerobe.2010.03.001.

Molloy, E. M., Collin Hill, Paul D. Cotter, and Paul Ross. "Bacteriocins." In *Reference Module in Food Science, from Encyclopedia of Dairy Sciences, 2nd*. 420–29. Elsevier Inc., 2011. https://doi.o Lim rg/10.1016/B978-0-12-374407-4.00049-2.

Mukai, R., Osamu Handa, Yosuke Suyama, Atsushi Majima, and Yuji Naito. "Effectiveness of including probiotics to *Helicobacter pylori* eradication therapies." *Journal of Clinical Biochemistry and Nutrition* 67(1), (2020):102–104. https://doi.org/10.3164/jcbn.20-37.

Myllyluoma, E., Lea Veijola, Tiina Ahlroos, Soile S.H. Tynkkynen, Esko Kankuri, Hansson I Vapaatalo, Hilpi Rautelin, and Riitta A Korpela. "Probiotic supplementation improves tolerance to *Helicobacter pylori* eradication therapy–a placebo-controlled, double-blind randomized pilot study." *Alimentary Pharmacology and Therapeutics* 21, (2005): 1263–1272. https://doi.org/10.1111/j.1365-2036.2005.02448.x.

Owen, R. J. "Bacteriology of *Helicobacter pylori*." *Bailliere's Clinical Gastroenterology* 9(3), (1995): 415–446. https://doi.org/10.1016/0950-3528(95)90041-1.

Pasolli, E., Francesca De Filippis, Italia E. Mauriello, Fabio Cumbo, Aaron M. Walsh, John Leech, Paul D. Cotter, Nicola Segata, and Danilo Ercolini. "Large-scale genome-wide analysis links lactic acid bacteria from food with the gut microbiome." *Nature communications* 11, (2020): 2610. https://doi.org/10.1038/s41467-020-16438-8.

Ponce, A. G., María R. Moreira, Carlos Enrique del Valle, and Sara Inés Roura. "Preliminary characterization of bacteriocin-like substances from lactic acid bacteria isolated from organic leafy vegetables." *LWT* 41, (2008): 432–441. https://doi.org/10.1016/j.lwt.2007.03.021.

Qin, J., Ruiqiang Li, Jeroen Raes, Manimozhiyan Arumugam, Kristoffer Solvsten Burgdorf, Chaysavanh Manichanh, et al. "A human gut microbial gene catalog established by metagenomic sequencing." *Nature* 464, (2010): 59–65. https://doi.org/10.1038/nature08821.

Reuter, G. "Composition of the microflora of human small intestine and the behaviour of some microorganisms after oral intake." *Proceedings of the 1st Intersectional Congress of International Association of Microbiological Societtes*, Tokyo, ed. By Hasegawa, T.: Science Council of Japan 2, (1975): 327–339.

Reuter, G. "Das Vorkommen von Laktobazillen in *Lactobacillus* and *Bifidobacterium* 53 Lebensmitteln und ihr Verhalten im menschlichen Intestinaltrakt." *Zentralblatt für Bakteriologie, Parsitenkunde, Infektionskrankheiten und Hygiene, Abteilung 1. Originale*, 1965a. 197, 468–487.

Reuter, G. "The *Lactobacillus* and *Bifidobacterium* microflora of the human intestine: composition and succession." *Current Issues in Intestinal Microbiology* 2(2), (2001): 43–53.

Reuter, G. "Untersuchungen über die Zusammensetzung und die Beeinflußbarkeit der menschlichen Magen- und Darmflora unter besonderer Berücksichtigung der Laktobazillen." *Ernährungsforschung* 1965b. 10: 429–435.

Schulz, C., Kerstin Schütte, Nadine Koch, Ramiro Vilchez-Vargas, Melissa L. Wos-Oxley, Andrew P. A. Oxley, Marius Vital, Peter Malfertheiner, and Dietmar H. Pieper. "The active bacterial assemblages of the upper GI tract in individuals with and without *Helicobacter* infection." *Gut* 67, 2018: 216-225. https://doi.org/10.1136/gutjnl-2016-312904.

Sheu, B. S., Jiunn-Jong Wu, C.-Y. Lo, H. W. Wu, Jyh-Hong Hong Chen, Yee-Shin Lin, and Ming T. Lin. "Impact of supplement with *Lactobacillus*- and *Bifidobacterium* containing yogurt on triple therapy for *Helicobacter pylori* eradication." *Alimentary Pharmacology and Therapeutics* 16, (2002): 1669–1675. https://doi.org/10.1046/j.1365-2036.2002.01335.x.

Simova, E. D., DoraBeshkova, and Zhechko P. Dimitrov. "Characterization and antimicrobial spectrum of bacteriocins produced by lactic acid bacteria isolated from traditional Bulgarian dairy products." *Journal of Applied Microbiology* 106, (2009): 692–701. https://doi.org/10.1111/j.1365-2672.2008.04052.x.

Štšepetova, J., Epp Sepp, Helgi Kolk, Krista Lõivukene, Epp Songisepp, and Marika Mikelsaar. "Diversity and metabolic impact of intestinal *Lactobacillus* species in healthy adults and the elderly." *British Journal of Nutrition* 105, (2011): 1235–1244. https://doi.org/10.1017/S0007114510004770.

Szajewska, H., Piotr Albrecht, and Agnieszka Topczewska-Cabanek. "Randomized, double-blind, placebo-controlled trial: effect of *Lactobacillus* GG supplementation on *Helicobacter pylori* eradication rates and side effects during treatment in children." *Journal of Pediatric Gastroenterology and Nutrition* 48, (2009): 431–436. https://doi.org/10.1097/mpg.0b013e318182e716.

Techo, S., Wonnop Visessanguan, Ratha-korn Vilaichone, Somboon Tanasupawat. "Characterization and antibacterial activity against *Helicobacter pylori* of lactic acid bacteria isolated from Thai fermented rice noodle." *Probiotics and Antimicrobial Proteins* 11, (2019): 92–102. https://doi: 10.1007/s12602-018-9385-z.

Thuy, T. T. D., Pei-Yun Kuo, Sha-Ma Lin and Cheng-Yen Kao. "Anti-*Helicobacter pylori* activity of potential probiotic *Lactiplantibacillus pentosus* SLC13." *BMC Microbiology* 22, (2022): 277. https://doi.org/10.1186/s12866-022-02701-z.

Tsai, C. C., Li-Fang Huang, Chia-Chen Lin, Hau-Yang Tsen. "Antagonistic activity against *Helicobacter pylori* infection *in vitro* by a strain of *Enterococcus faecium* TM39." *International Journal of Food Microbiology* 96, (2004): 1–12. https://doi.org/10.1016/j.ijfoodmicro.2003.10.019.

Vaughan, E. E., Hans G. H. J. Heilig, Kaouther Ben-Amor, and Willem M. de Vos "Diversity, vitality and activities of intestinal lactic acid bacteria and bifidobacteria assessed by molecular approaches." *FEMS Microbiology Reviews* 29(3), (2005): 477–90. https://doi.org/10.1016/j.femsre.2005.04.009.

Viazis, N., Konstantinos Argyriou, Katerina Kotzampassi, Dimitrios K Christodoulou, Periklis Apostolopoulos, Sotirios D. Georgopoulos, et al., "A four-probiotics regimen combined with a standard *Helicobacter pylori*-eradication treatment reduces side effects and increases eradication rates." *Nutrients* 14, 2022, 632. https://doi.org/10.3390/nu14030632.

Walter, Jens. "Ecological role of Lactobacilli in the gastrointestinal tract: implications for fundamental and biomedical research." *Applied and Environmental Microbiology* 74(16), (2008): 4985–4996. https://doi.org/10.1128/AEM.00753-08.

Wang, K. Y., Shui-Nin Li, Chiang-Shin Liu, Daw-Shyong Perng, Yu-Chung Su, Deng-Chyang Wu, Chang-Ming Jan, Chun-Huang Lai, Tsu-Nai Wang, and Wen-Ming Wang. "Effects of ingesting *Lactobacillus*- and *Bifidobacterium*-containing yogurt in subjects with colonized *Helicobacter pylori*." *The American Journal of Clinical Nutrition* 80, (2004): 737–741. https://doi.org/10.1093/ajcn/80.3.737.

Wang, M., Siv Ahrné, Bengt Jeppsson, and Göran Molin. "Comparison of bacterial diversity along the human intestinal tract by direct cloning and sequencing of 16S rRNA genes." *FEMS Microbiology Ecology* 54, (2005): 219–231. https://doi.org/10.1016/j.femsec.2005.03.012.

Woraprayote, W., Laphaslada Pumpuang, Amonlaya Tosukhowong, Sittiruk Roytrakul, Rodney Honrada Perez, Takeshi Zendo, Kenji Sonomoto, Soottawat Benjakul, Wonnop Visessanguan. "Two putatively novel bacteriocins active against Gram-negative food borne pathogens produced by *Weissella hellenica* BCC 7293." *Food Control* 55, (2015): 176–184. https://doi.org/10.1016/j.foodcont.2015.02.036.

Woraprayote, W., Thitiphorn Janyaphisan, Saowalak Adunphatcharaphon, Nipon Sonhom, Kittiya Showpanish, Kittaporn Rumjuankiat, Wonnop Visessanguan, Christopher T. Elliott, Awanwee Petchkongkaew. "Bacteriocinogenic lactic acid bacteria from Thai fermented foods: Potential food applications." *Food Bioscience* 52, (2023): 102385. https://doi.org/10.1016/j.fbio.2023.102385.

Yap, T. W. C., Han-Ming Gan, Yin-Peng Lee, Alex Hwong-Ruey Leow, Ahmad Najib Azmi, Fritz Francois, et al., "*Helicobacter pylori* eradication causes perturbation of the human gut microbiome in young adults." *PLoS ONE* 11(3), (2016): e0151893. https://doi.org/10.1371/journal.pone.0151893.

Yin, Y. N., Chun-Lian Wang, Xiao-Wei Liu, Yi Cui, Ning Xie, Qiong-Fen Yu, Fu-Jun Li, Fang-Gen Lu. "Gastric and duodenum microflora analysis after long-term *Helicobacter pylori* infection in Mongolian Gerbils." *Helicobacter* 16(5), (2011): 389–97. https://doi.org/10.1111/j.1523-5378.2011.00862.x

Zendo, T. "Screening and characterization of novel bacteriocin from lactic acid bacteria." *Bioscience, Biotechnology, and Biochemistry* 77(5), (2013): 893–899. https://doi.org/10.1271/bbb.130014.

Chapter 4

Phytoremediation Potential of *Azolla* Species: A Comprehensive Review of Physiological, Biochemical, and Molecular Responses to Heavy Metal Contamination in Freshwater Environments

Seran Kirkinci[1]
Yusuf Can Gercek[2]
and Taylan Kosesakal[2,*]
[1]Institute of Graduate Studies in Sciences, Istanbul University, Suleymaniye, Istanbul, Turkey
[2]Botany Division, Department of Biology, Faculty of Science, Istanbul University, Vezneciler, Istanbul, Turkey

Abstract

The escalating trends of industrialization and urbanization have precipitated the swift contamination of vital freshwater reservoirs by environmental pollution. Inorganic chemicals, notably heavy metals (HMs), give rise to a spectrum of environmental and economic issues. Furthermore, these contaminants exert deleterious effects on soil and water quality, as well as on the nutritional dynamics of plants and animals, consequently impacting human health. HMs such as zinc (Zn), nickel (Ni), copper (Cu), cadmium (Cd), lead (Pb), arsenic (As), and chromium (Cr) rank among the most pervasive water pollutants, posing a significant menace to the environment. The indispensability of water

* Corresponding Author's Email: taylank@istanbul.edu.tr.

In: Advances in Biology. Volume 6
Editor: Charles D. Grant
ISBN: 979-8-89113-712-7

for human sustenance, coupled with an increasing global awareness of environmental matters, especially within aquatic ecosystems, has spurred researchers worldwide to focus on the development of cost-effective and environmentally friendly technologies for the removal of both organic and inorganic pollutants, thereby enhancing water quality.
In this context, phytoremediation emerges as a promising and sustainable technology for the removal of inorganic pollutants from freshwater sources. However, the efficacy of phytoremediation varies among plant species due to differences in growth patterns, biomass production, developmental stages, and pollution levels. Among aquatic plants, a free-floating fern, *Azolla*, characterized by its rapid growth, and nitrogen-fixing properties, stands out as a compelling candidate for the extraction, disposal, and recovery of pollutants from aquatic ecosystems. Azolla species demonstrate a notable capacity for phytoremediation and bioconcentration in media contaminated with HMs. Nevertheless, it is crucial to acknowledge that HM concentrations exert morphological, physiological, and molecular effects on plants.
This review critically assesses the physiological, biochemical, and molecular impacts of HMs contamination on *Azolla* species, shedding light on the potential application of this plant for phytoremediation purposes.

Keywords: *Azolla*, heavy metal, phytoremediation, tolerance mechanism

Introduction

Pollutants are substances introduced into the environment that can cause harm to living organisms and ecosystems. These substances often originate from human activities such as industrial processes, transportation, agriculture, and waste disposal. Common pollutants include HMs, pesticides, industrial chemicals, and air pollutants such as particulate matter and gases. Pollutants can have various adverse effects on human health, wildlife, and the environment, including respiratory problems, reproductive issues, and ecosystem disruption. Understanding the sources, distribution, and impact of pollutants is crucial for effective environmental management and protection (Giller, Witter, and Mcgrath, 1998; Manisalidis et al., 2020). Pollutants are found in soil, water, and air. Aquatic environments, especially freshwater sources have been exposed to serious threats from pollutants (Amoatey and Baawain, 2019). One of the abiotic stress factors is HMs. Although the toxicity of HMs depends on their concentration, toxicity is also influenced by their chemical and physical configurations. The mobility and bioavailability of

HMs affect their behavior in freshwater and determine their toxic effects on organisms (De Paiva Magalhães et al., 2015).

HMs have been named in life sciences as elements, that show toxic effects on the organisms' metabolic systems, when a certain HM concentration in the environment is exceeded (Appenroth, 2010). HMs are one of the major causes of environmental pollution. HMs pollution is caused by various metals, especially copper (Cu), nickel (Ni), cadmium (Cd), zinc (Zn), chromium (Cr) and lead (Pb). Some HMs (like iron (Fe), Zn, calcium (Ca) and magnesium (Mg)) have been reported to be essential to many organisms. However, some others (like arsenic (As), Cd, Pb, and methylated forms of mercury (Hg)) are not essential for organisms and they show toxic effects on organism metabolite mechanism (Sing and Kalamdhad, 2011). Non-essential elements are toxic to organisms in their ionic and methylated forms. When they enter into the biological body, they can damage various cellular mechanisms through their strong interaction with thiol. The reactivity of thiols can inhibit protein function and lead to oxidative stress (Clemens and Ma, 2016). Plants have evolved strategies for the uptake, transport, and accumulation of HMs. Their behavior depends on the chemical features present in the aquatic environment and the interactions with anions and molecules within plant cells (Inouhe et al., 2012).

Effects of HMs on Plants

The uptake of HMs into plant cells is regulated through symplastic and apoplastic pathways. In the apoplastic way, HMs are moved into intercellular space; however, in the symplastic way, they are translocated cell to cell by membrane transport proteins, including the ZIP-family, P-type ATPase-HMA, MATE family, OPT family (Verbruggen, Hermans and Schat, 2009). Heavy metal uptake by plant roots and their transport between cells is shown in Figure 1. It has been recognized that some can grow, reproduce as well as survive on sites heavily polluted with HMs. The types of species that can tolerate high concentrations of HMs are intensely toxic to the other species' metabolism. They are called hyperaccumulators that develop tolerance or in some cases hypertolerance. Hyperaccumulators and non-hyperaccumulators detoxify HMs by chelation in the cytoplasm or by storing them in the vacuole, then translocate towards the shoot via the xylem. One of the tolerance strategies is that HMs' detoxification/sequestratin occur such as epidermis where the cells are located away from photosynthetic cells for less damage to the photosynthesis mechanism. Non-hyperaccumulators rely on more phytochelatins than hyperaccumulators and overexpress the antioxidant-related genes for detoxifying the plant cells with HMs accumulations (Schat et al., 2002; Rascio and Navari-Izzo, 2011). Some plants survive the toxic effects of HMs through tolerance mechanisms and depending on the concentration levels of HMs. Previous studies indicate that plants have developed tolerance strategies by overexpressing antioxidant activity and detoxifying HMs. However, the growth and development of plants are affected by exposure to HMs (Breckle and Kahle, 1992; Wani, Khan, and Zaidi, 2008; Imran et al., 2013; Fahad et al., 2019).

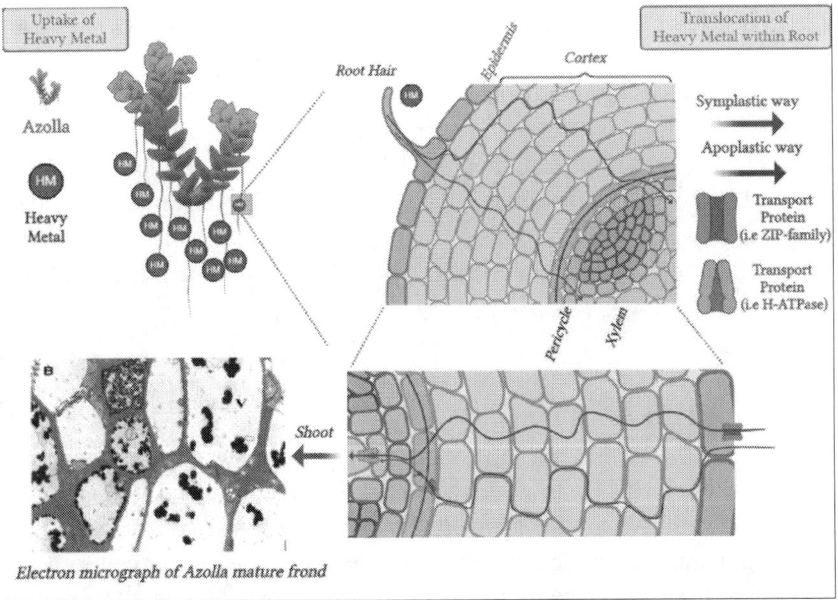

Figure 1. HMs are transported in root cells by symplastic and apoplastic pathways after uptake by plant root hairs. In the symplastic pathway, it is transported along the plasmodesmata and passes to the xylem using transport proteins. The figure shows a root cross-section of a dicotyl plant root system. In *Azolla*, HMs uptake by root hairs translocate through the root system as described above and accumulate in the shoot. The black clustered shapes in the cross-section of the *Azolla* mature frond show Pb accumulated in the vacuole (Benaroya et al., 2004). V: vacuole. The figure was created with www.BioRender.com. All rights and ownership of BioRender content are reserved by BioRender. BioRender content included in the completed graphic is not licensed for any commercial uses beyond publication in a book.

One of the detoxification/sequestration strategies of HMs is phytochelates (PCs), which are short-chain cysteine-rich peptides with -SH groups that can bind to metals. PCs are often involved in the intracellular transport and retention of HMs such as Cd, As, Pb, Zn, and Cu ions in vacuoles. They are responsible not only for the hyperaccumulation mechanism of metals but also for metal homeostasis (Seregin and Kozhevnikova, 2023). PCs contain only three amino acids (L-cysteine, L-glutamic acid, and glycine). In general, their structure is (γ-Glu-Cys)n-Gly (n = 2-11) (Gekeler et al., 1989; Chaudhary, Agarwal, and Khan, 2018). Metallothioneins (MTs) are short-chain peptides from the phytochelatin group that binds to metals via thiol groups of cysteine and are divided into four groups (MT1, MT2, MT3, and MT4). Several transport protein families are involved in the transport of HMs into cell compartment. They are involve in the detoxification/sequestration process. Overexpression of PC and MT genes leads to increased expression of HMs ATPase genes (HMAs), an integral membrane transport protein involved in the uptake of transition metals. In addition, organic acids and amino acids are also associated with the detoxification process (Tomsett and Thurman, 1988; Hassinen et al., 2010; Sign et al., 2016).

After the uptake of HMs into plant cells, it has been recognized that metal ions tend to stick to the sulfhydryl groups of proteins and the hydroxyl groups of phospholipids (Prasad, Hagemeyer, and Devi, 1999). Additionally, they can displace calcium ions from critical locations on cell membranes (Breckle, 1991). It has also been reported many times that high concentrations of HMs, especially Cu and Cd, decrease the activity of plasma membrane H+-ATPase (Fodor, Szabó-Nagy and Edei, 1995; Burzyński and Kolano, 2003; Janicka-Russak et al., 2008), an enzyme found in plant and fungal plasma membranes that plays a crucial role in regulating the transport of protons across the membrane, maintaining the pH and ionic balance of cells by pumping hydrogen ions out of the cell. Accumulation of high concentrations of HMs can disrupt the cell membrane's integrity as well as the cells' ionic balance; moreover, accumulation of high concentrations of HMs can lead to the overproduction of reactive oxygen species (ROS), including superoxide (O_2^-), hydrogen peroxide (H_2O_2), hydroxyl radical (OH), singlet oxygen ($^1O_2$), and hydroperoxides (ROOH) within plants' cells. Localizations and reactions of ROS production in the cell are shown in Figure 2. In contrast to redox-active metals such as Fe and Cu, HMs like Pb, Cd, Ni, Al, Mn, and Zn are unable to generate ROS directly through biological redox reactions such as Haber Weiss/Fenton reactions. Nevertheless, these metals can trigger ROS generation through various indirect mechanisms, such as the stimulation of

NADPH oxidases' activity, the displacement of crucial cations from specific enzyme binding sites, and the inhibition of enzymatic activities due to their affinity for -SH groups on the enzyme (Shahid et al., 2014). Typically, ROS serves as a crucial regulator of gene expression, influencing a variety of processes such as cellular growth, plant development, stress response, signaling, cell death, and defense against pathogens. Nonetheless, if ROS production exceeds the capacity of antioxidants to neutralize them, oxidative stress may arise, damaging plant physiology (Mansoor et al., 2023). One of the most important interactions caused by ROS is lipid peroxidation, which arises when a ROS such as .OH or O_2- comes into contact with a lipid molecule, usually a polyunsaturated fatty acid (PUFA). The process of lipid peroxidation results in the breakdown of lipid components found in cell membranes. This breakdown can disrupt the structure of the lipid bilayer, leading to membrane leakage (Catalá 2016). Lipid peroxidation generates reactive aldehydes such as malondialdehyde (MDA) and 4-hydroxynonenal (4-HNE). These aldehydes can covalently modify proteins and other biomolecules in the membrane, altering their structure and function. This modification can disrupt membrane proteins involved in maintaining membrane integrity and regulating permeability, leading to increased permeability of the membrane (Ayala et al., 2014). The measurement of electrolyte leakage is often used as a marker of membrane damage caused by HMs. (Jambunathan, 2010).

Plants have developed various defense and tolerance mechanisms against environmental factors and their stressors. Antioxidants are classified as enzymatic (including SOD, CAT, POD, APX, MDHAR, DHAR, GR and GPX) and non-enzymatic (including AA, GSH, α-tocopherol, carotenoids, flavonoids, and proline). The localization of the production of antioxidants and the reactions in which they participate are shown in Figure 3. In the plant cell, their production increases to develop plant defense mechanisms and tolerance against stress factors such as HMs stress. (Das and Roychoudhury, 2014). Enzymatic antioxidants are involved in the major ROS scavenging pathways of plants. The balance between SOD and APX is important for measuring the activity of O_2^- and H_2O_2 and for sequestering metal ions to prevent the formation of highly toxic -OH radicals. CAT degrades H_2O_2 and is exclusively found in peroxisomes. GSH is a major enzyme involved in scavenging and is also involved in phytochelatin biosynthesis. The chelation of HM ions by GSH indicates that expressing GSH molecules play a critical role in the detoxification process (Bhaduri and Fulekar, 2012). Anthocyanins are molecules belonging to the flavonoid family that give natural color to flowers, fruits, roots and leaves. Under abiotic stress, their expression increases for creating tolerance to the plant due to increased ROS production (Naing and Kim, 2021).

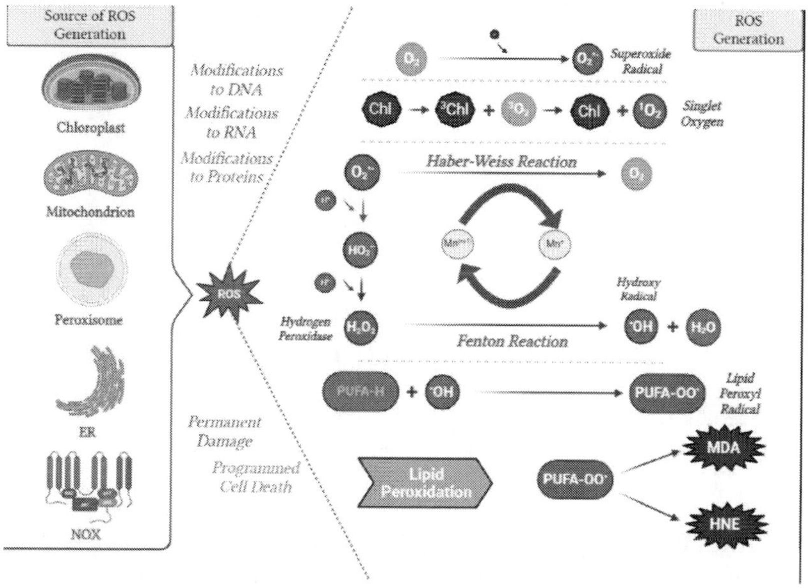

Figure 2. ROS accumulation in plants depends on the balance between ROS production and ROS scavenging. In addition, environmental conditions that generate stressors such as HMs can upset this balance. ROS are unstable, highly reactive, and act by reacting with other macromolecules to form more free radicals. As a result of these interactions, modifications occur in DNA, RNA, and proteins and unfavorable conditions are created that can lead to cell death. The NADPH oxidase protein in cell membranes is the main source of ROS. In addition, ROS is produced in the chloroplast, mitochondria, ER, and peroxisome organelles. ROS production reactions are shown in the figure above. The Haber-Weiss and Fenton reactions are expressed as an example from a study by Ayala, Muñoz, and Argüelles, (2014). ER: Endoplasmic reticulum NOX: NADPH oxidase, O_2: oxygen, Chl: Chlorophyll, $_3$Chl: triplet excited state of chlorophyll, H_2O: water, $HO_2^{\cdot -}$: hydroperoxyl, PUFA-H: polyunsaturated fatty acid, MDA: 3,4-Methylene dioxy-amphetamine, HNE: 4-hydroxynonenal. The figure was created with www.BioRender.com. All rights and ownership of BioRender content are reserved by BioRender. BioRender content included in the completed graphic is not licensed for any commercial uses beyond publication in a book.

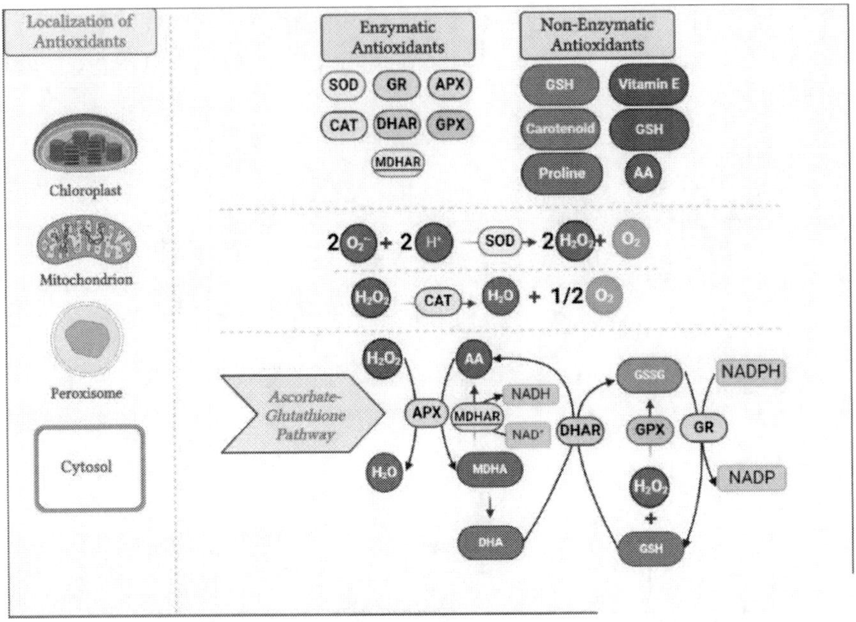

Figure 3. Antioxidants are compounds that protect cells against damage caused by highly reactive ROS. They induce reactions that inhibit the harmful effects of ROS in many compartments of the cell such as chloroplast, mitochondria, peroxisomes, and cytosol. The reduction reactions of antioxidants are expressed in the figure above. The Ascorbate-Glutathione Pathway is expressed as an example from a study by Pandey et al., (2015). The pathway starts with the reduction of MDHA to AA by MDHAR enzyme or DHA to AA by DHAR enzyme and detoxifies H_2O_2 via APX enzyme. The reduction of the DHA molecule to AA occurs with the use of GSH as a reductant: GSH is again oxidized from GSSG to GSH by the NADPH-dependent enzyme GR. SOD: superoxide dismutase, CAT: catalase, GR: glutathione reductase, APX: ascorbate peroxidase, DHAR: dehydroascorbate reductase, GPX: guaiacol peroxidase, MHDAR: monodehydroascorbate reductase, GSH: glutathione, AA: ascorbate, MDHA: monodehydroascorbate, DHA: dehydroascorbate, GSSG: oxidized glutathione, NADPH/NADP: Nicotinamide adenine dinucleotide phosphate, NADH/NAD^+: Nicotinamide adenine dinucleotide, $O_2^{\cdot -}$: superoxide radical, H^+: hydrogen peroxidase, H_2O_2: hydrogen peroxide, O_2: oxygen, H_2O: water. The figure was created with www.BioRender.com. All rights and ownership of BioRender content are reserved by BioRender. BioRender content included in the completed graphic is not licensed for any commercial uses beyond publication in a book.

Phytoremediation Potential of *Azolla*

Water is indispensable for human nutrition, and increasing awareness of environmental issues, particularly aquatic ecosystems, has drawn the interest of researchers worldwide. In this regard, there is a need for low-cost and environmentally friendly technologies to remove both organic and inorganic pollutants from the environment and thus improve water quality. Phytoremediation presents an attractive alternative (Sood et al., 2012). As an emerging technology, phytoremediation is the process of removing, degrading, or sequestering chemical pollutants from soil, sediment, groundwater, surface water, and even the atmosphere using plants and associated rhizospheric microorganisms (Chappell, 1997). The removal of pollutants from the environment is achieved through different phytoremediation mechanisms depending on the type of pollutant. Generally, phytoremediation mechanisms can be categorized under five headings: Phytoextraction, phytovolatilization, phytodegradation, rhizofiltration, and phytostabilization (Table 1).

Phytoremediation emerges as a low-cost and eco-friendly remediation technology for the removal of organic and inorganic pollutants (Yan et al., 2011; Rezania et al., 2016). Phytoremediation is a favorable green technology in polluted aquatic environments such as wastewater remediation by using macrophytes and microorganisms to remove, stabilize or degrade the HMs (Pang et al., 2023). Schwitzguébel (2001), identified that various plant species can accumulate toxic metals in their tissues. Previous studies showed that some of free-floating macrophytes, including, duckweed, water lettuce, and mosquito fern, have eliminated various pollutants such as HMs (Ali et al., 2020). The research showed that selected macrophytes such as *Lemna*, *Salvina*, and *Azolla*, remove HMs from polluted freshwater (Mishra et al., 2008; Rai, 2009a; Rai, 2009b; Vaseem and Banerjee, 2012; Bharti and Banerjee, 2012; Kumari, Lal and Rai, 2016; Xinwei et al., 2018; Amare et al., 2018; Banach et al., 2020). However, plant species vary in their phytoremediation potentials depending on factors such as plant growth, biomass production, developmental stage, and pollution levels (Williams, 2002). Among aquatic plants, *Azolla*, a free-floating, fast-growing fern capable of nitrogen fixation, appears to be an excellent candidate for the removal, remediation, and recovery of pollutants from water ecosystems. *Azolla*, lives in symbiosis with the nitrogen-fixing cyanobacterium '*Nostoc azollae* 0708' (Ran et al., 2010). Belonging to the Azollaceae family and having seven different species worldwide, *Azolla* is typically found floating on the surface of lakes, ponds, wetlands, channels, rice fields, and slow-flowing rivers in tropical and temperate regions (van Hove and Lejeune, 2002). *Azolla*, holds great importance for agricultural activities in both developed and developing countries. The *Azolla - N. azollae* symbiosis is the only symbiotic relationship between the eukaryotic partner *Azolla* and the prokaryotic endosymbiont *N. azollae*. The agronomic potential of this relationship stems from the successful growth of the plant in nitrogen-deficient or low-nitrogen areas (Carrapiço, 2010). Due to its nitrogen-fixing symbiotic partner living in dorsal leaf cavities, *Azolla* is often used as a fertilizer in rice fields. Successful results have been achieved in environmental management through the use of *Azolla*, including the control of weeds and mosquitoes and the removal of organic and inorganic pollutants from freshwater bodies, owing to its high growth rates and phytoremediation potentials (Mashkani and Ghazvini, 2009; Forni et al., 2012; van Kempen et al., 2013; Akinbile et al., 2016; Kosesakal et al., 2016). Under optimal growth conditions, *Azolla* can double its biomass in approximately 2-4 days. This rapid biomass increase makes *Azolla* an ideal candidate for use in phytoremediation systems (Roberts et al., 2014). This freshwater fern, which holds biotechnological value due to its symbiotic relationship with the cyanobacterium *N. azollae*, possesses high biomass productivity and a remarkable capacity for environmental remediation (Bennicelli et al., 2004). Furthermore, its ease of harvest and drying makes it an ideal candidate for phytoremediation studies in both water and wastewater (Arora et al., 2006). Therefore, *Azolla* is a good bioremediator that can be used for the treatment of industrial wastewater and sewage, being both environmentally friendly and effective (Talebi et al., 2019). *Azolla* species exhibit high phytoremediation and bioconcentration potentials in growth environments contaminated with HMs. However, high concentrations of HMs can also cause morphological, physiological, biochemical, and molecular effects on plants. The physiological and molecular effects of different heavy metals on *Azolla* species are summarized in Table 2 and Table 3, respectively.

Table 1. Phytoremediation mechanisms

Phytoremediation Mechanisms	
Phytoextraction	Plants accumlate pollutants and store them in harvestable plant parts such as root, steams and leaves.
Phytostabilization	Plants restrict the mobiity and bioavailability of pollutants in the root zone through immobilization or precipitation.
Phytodegradation	Plants reduce organic pollutants using hyrolytic enymes and metabolites.
Rhizofiltration	Plant roots adsorb, accumulate or precipitate pollutants from wastewater.
Phytovolatilization	Conversion of pollutants to volatile form and their subsequent release to the atmosphere.

Table 2. Physiological effects of HMs on *Azolla* species

Species	**HMs**	**Initial concentration of HMs**	**Duration of experiment (days)**	**Physiology**	**References**
A. pinnata	Hg	5-30% soil waste	30 and 60 days	Decreased on fresh weight has been observed at increasing concentrations.	(Mishra, Nanda and Misra, 1987)
A. filiculoides	Cu, Cd and U	10 ppm	1 (Cu), 2 (Cd) and 4 (U) days	The potassium (K) and magnesium (Mg) content has decreased on the roots. Despite the decrease in calcium (Ca) content under the influence of cadmium (Cd) and copper (Cu), it has increased in the presence of uranium (U).	(Sela et al., 1988)
A. pinnata	Cd, Cr, Co, Cu, Ni, Pb and Zn	0.75-75 μM Zn; 0.24-96 μM Pb; 0.85-85 μM Ni; 0.75-80 μM Cu; 0.95-192 μM Cr; 0.85-85 μM Co; 0.45-45 μM Cd.	4 days	Inhibition in growth	(Gaur, Noraho abd Chauhan, 1994)
A. caroliniana	Hg	0.1, 0.5 and 1.0 mg dm-3	12 days	Inhibition in growth	(Bennicelli et al., 2004)
A. pinnata	Hg	10 μg L-1	28 days	Decreased on chlorophyll and protein content	(Mishra, Tripathi and Kim, 2009)

Table 2. (Continued)

Species	HMs	Initial concentration of HMs	Duration of experiment (days)	Physiology	References
A. imbricata	Cd	0.01, 0.05, 0.1, and 0.5 mg L-1	1-9 days	Cd could induce color change on the surface of the fronds, reduced chlorophyll and carotenoid content, increased anthocyanin and total phenolic content, and PAL activty	(Dai et al., 2006)
A. pinnata	Hg and Cd	0.5, 1 and 3 mg L-1	13 days	Inhibition in growth	(Rai, 2008)
A. filiculoides	Cs and Sr	100 mg L-1	15 days	Inhibition in growth	(Mashkani and Ghazvini, 2009)
A. pinnata	Hg	0.1 , 0.5, 1 and 3 mg L-1	6 days	Decreased on chlorophyll-a and protein content.	(Rai and Tripathi, 2009)
A. caroliniana	Pb	1. 5, 10, and 20 mg L-1	10 days	Decreased on some minerals (Ca, Mg, Zn and N) and chlorophyll, as well as carotenoid content	(Roberts, Boylen and Nierzwicki-Bauer, 2014)
A. caroliniana	As, Si and Cu	0-20 µM (As), 2-78 µM (Cu), and 0-1.8 mM (Si)	14 days	Inhibition in growth. Increase in anthocyanin content in presence of As and combination.	(Rofkar, Dwyer and Bobak, 2014)
A. pinnata	Cd	0.05, 0.1, 0.5, 1.0, 1.5 and 2.0 mg L-1	1 and 4 days	At high concentrations, despite a decreaseon biomass and chlorophyll content, there has been an increase in carotenoid content, photosynthesis and photorespiration rates, as well as antioxidant activity.	(Prasad and Singh, 2011)
A. filiculoides	As	5, 10, 20 and 30 µg L-1	4 days	Caused on significant reductions in the contents of Mn, Fe, Zn, S, Ca, and K.	(Sánchez-Viveros et al., 2016)
A. imbricata	Cd	0.1 mg L-1	1 day	By increasing the anthocyanin content, antioxidant activity has increased.	(Dai, Dong and Ma, 2012a)
A. pinnata	Hg, As, Pb, Cu, Cd and Cr	1, 2, 5 mg L-1	28 days	Decreased on Hill acitivity, chlorophyll and protein content, as well as dry weight, but increased on membrane permeability	(Sarkar and Jana, 1986)
A. filiculoides	Cs and Sr	100 mg L-1	15 days	Inhibition in growth	Mashkani and Ghazvini, 2009

Species	HMs	Initial concentration of HMs	Duration of experiment (days)	Physiology	References
A.microphylla	Cr	1, 5, 10, 15 and 20 ppm	7 days	Decreased fresh weight. At high concentration, depressed nitrogenase activity	(Arora et al., 2006)
A. pinnata					
A. filiculoides					
A. caroliniana	Pb (II)	0.1 mg dm-3	12 days	Decreased on biomass content	(Stepniewska et al., 2005)
		0.5 mg dm-3			
		1 mg dm-3			
	Cd (II)	0.1 mg dm-3			
		0.5 mg dm-3			
		1 mg dm-3			
A. filiculoides	As	5, 10, 20, 30, 60 and 120 μg mL-1	4 days	At high concentration, decreased on growth and chlorophyll and total phenolic content, but significant increased on total carotenoid+xanthophylls content.	(Sánchez-Viveros, Ferrera-Cerrato and Alarcón, 2011)
A. filiculoides	Cu	0.02, 0.2, 1 and 2 mM	36 h	Caused on membrane damage. At 0.02 mM concentration, increased on phenolic content	(Sánchez-Viveros et al., 2010)
A. caroliniana					

Table 3. Gene expressions of *Azolla* species under HMs stress

Species	**HMs**	**Gene/Protein**	**References**
A. filiculoides	Pb	V-H+-ATPase	(Benaroya et al., 2004)
A. filiculoides	Cd, Cu, Zn and Ni	*AzMT2*	(Schor-Fumbarov et al., 2005)
A. imbricata	Cd	*CHS* and *DFR*	(Dai, Dong and Ma, 2012b)
A. pinnata	Cu, Cd, Zn and Ni	*MT2* and *PCS1*	(Talebi et al., 2019)
Anzali			
A. filiculoides			

Cd

Cd is one of the HMs that has a negative impact on human health and is not essential for plants. It is often released into the environment due to industrial activities such as metal mining, zinc refining, and intensive use of herbicides and fertilizers (Song, Jin and Wang, 2017). Previous studies have found that both terrestrial plants and macrophytes develop tolerance mechanisms to Cd stress at certain concentrations (Citterio et al., 2003; Clabeaux et al., 2011; Nie et al., 2016; Romero-Hernández et al., 2017). In terrestrial and aquatic environments, Cd accumulation occurs firstly on the root hairs and then the epidermal cell, secondly Cd is transported across the root cortex, entering the conductive vessels of the xylem. After passing through all the barriers from the root, including the apoplasmic barriers, the metal ions enter the symplast and are transported to the stele and the xylem elements (Akhter et al., 2014). Previous studies have identified that many transport proteins are used to move Cd (NRAMP family, P-type ATPase, ABC transporter, CAX family, ZIP family, LCT transporter, and CE family). After Cd is taken up into the plant by the roots, it is sequestered in inactive tissues to keep it away from the active tissues (Ismael et al., 2019; Mitra et al., 2014).

As a result of Cd absorption from *Azolla* roots, it was found to be localized intensively both on the outer wall of the root epidermis and on the cell wall of the bundle sheath xylem cell by transport via the xylem (Sela et al., 1988). The localization of Cd on the plant body varies depending on the concentration of Cd and the exposure period. In the early stages of exposure, Cd localization is higher in the roots (Sela et al., 1990). In a study with *Azolla* species, Cd accumulation increases under Zn deficiency compared to the Zn-containing test group. In addition, the use of Ca(II) channel blockers decreases Cd accumulation. The effect of Ca(II) channels and Zn transporters was observed in the regulation of Cd accumulation in *Azolla* (Tan et al., 2011). The expression of some PC biosynthesis genes, like *AzMT2* and PCS1, increase after Cd accumulation in *Azolla* (Schor-Fumbarov et al., 2005, Talebi, Tabatabaei and Akbarzadeh, 2019). Cd, depending on its concentration, Cd has toxic effects that create stress factors. Growth rate, intracellular ion homeostasis (Ca^{2+}, Mg^{2+}, Na^{+}, and K^{+}), photosynthetic efficiency, and antioxidant activity were decreased depending on the Cd concentration and the exposure period (Sela, Garty, and Tel-Or, 1989; Gaur, Noraha and Chauhan, 1994; Krupa, 1999; Stepniewska et al., 2005; Rai et al., 2008; Vaseem and Benerjee, 2012; Bharti and Banerjee, 2012; Valderrama et al., 2013). High concentrations of Cd (Sarkar and Jana, 1986) and its combinations with other HMs in the *A. pinnata* test solution had a toxic effect on the plant's body, reducing membrane integrity, hill activity, chlorophyll and protein content, and dry weight. In combination with Hg, cell permeability increased and the ion balance of the cell was negatively affected (Sarkar and Jana, 1987). The toxic effect of Cd increased MDA and H_2O_2 content to act as a ROS indicator and the content of enzymatic and non-enzymatic antioxidants. It was also observed in *Azolla* that the content of anthocyanins increased with Cd accumulation (Prasad and Singh, 2011). The expression of *CHS*, *DFR*, and PAL enzyme genes, which are among the genes responsible for anthocyanin biosynthesis, increased under HM stress in *Azolla* (Dai, Xiong and Li, 2006; Dai, Dong, and Ma, 2012b). During the exposure period when the concentration did not reach a fatal level for *Azolla*, the accumulation of Cd in the plant takes up with increasing concentration. The rate of accumulation is higher in the early stages of exposure and their BCF values were above 1 (Stepniewska et al., 2005; Rai, 2008; Dai et al., 2006; Naghipour et al., 2018; Amare et al., 2018). In addition to its use as a biofertilizer, *Azolla* has been investigated for its ability to reduce Cd accumulation in commercial crops such as rice and wheat due to its phytoremediation potential (Liu et al., 2021; Shedeed and Farahat, 2023).

Pb

Pb, which is not an essential element for living organisms, is released into the environment from various sources such as automobile exhaust and battery production processes. As with Cd, it moves through apoplastic and symplastic pathways into plants. In the symplastic pathway, Pb uses the P-type transport proteins, including ABC-transport proteins (Kumar and Prasad, 2018). In *Azolla* under Pb toxicity, the Pb content in apoplasts was found to be low. Microscopic studies revealed that Pb is accumulated in clustered form in vacuoles, as well as the cell walls of the leaf cell. As the Pb concentration increased, while the activity of the vacuolar H+ pump protein V-H+-ATPase was enhanced, its content was decreased (Benaroya et al., 2004). In a study with two different macrophytes, *A. microphylla* was found to accumulate more Pb in comparison to *P. stratiotes*. BCF and TF values of *A. microphylla* rose with increasing Pb concentrations (Rachmadiarti et al., 2018). It was determined that Pb has a less toxic effect on *Azolla* compared to Cd depending on the concentration and the exposure period. In a study on the absorption of various HMs by *Spirodela polyrhiza* and *A. pinnata* plants, it was determined that Pb showed less toxicity in both organisms compared to other HMs, and accordingly, the EC_{50} value for RGR was higher compared to other HMs except Zn. As with Cd, increasing concentrations of Pb increased its accumulation in *A. pinnata* biomass (Gaur, Noraho, and Chauhan, 1994; Naghipour et al., 2018). In the presence of Pb, photosynthesis and transpiration activities are inhibited by decreasing CO_2 fixation through stomatal closure (Parys et al., 1998). Plant chlorophyll-a, chlorophyll-b, carotenoid and anthocyanin contents decreased with increasing Pb concentrations. Accordingly, Mg and N contents decreased with increasing concentrations of Pb. The reason for the decrease in N content is the decrease in the frequency of heterocysts containing the nitrogenase enzyme of *Azolla* symbiont at increasing concentrations (Roberts et al., 2014; Stepniewska et al., 2005; Sharma and Sachdeva, 2015). Due to the toxic effects of Pb, plants have developed various tolerance mechanisms such as the production of antioxidant enzymes and non-enzymatic antioxidants (Kumar and Prasad, 2018). The toxic effect of Pb on *Azolla* varies depending on its concentration and exposure.

As

As is a naturally occurring metalloid that is not essential for plants. It is taken into the plant body by phosphate transport proteins (Farooq et al., 2016). Therefore, there is a competition between As and P. P content in the test solution increased as a result of As accumulation in *Azolla*. As has a negative effect on plant growth and development due to its toxic effect. When applied to *Azolla*, As decreases RGR, biomass, ion uptake and photosynthetic efficiency depending on the concentration and the exposure time (Sarkar and Jana, 1985; Sánchez-Viveros et al., 2016; Rofkar, Dwyer, and Bobak, 2014; Guimaraes et al., 2012). It was found that As accumulation in *A. caroliniana* decreased as a result of various combinations of As with other metals. The accumulation level of As may be reduced by utilizing the same transport proteins as many HMs (Gusman et al., 2013). The uptake and accumulation of As in the plant increases ROS production by affecting gas exchange and pigment content (Bali and Sidhu, 2021). *Azolla* developed a tolerance mechanism under the As application, moreover increasing its antioxidant activity (Sánchez-Viveros, Ferrera-Cerrato, and Alarcón, 2011; Gusman et al., 2013; Iriel et al., 2015; Leão et al., 2017). The microbiota, the *Azolla* symbiont, contributes to the development of plant tolerance mechanisms under

HM stress. Although plant biomass and chlorophyll content decreased in *A. microphylla* in combination with a bacterial colony, which is resistant to HM stress, the colony group showed positive effects on plant physiology compared to the only As-treated group. In addition to chlorophyll content, total carotenoid content was decreased by As. This may be due to the reduction of the photosynthetic capacity of the plant due to the inhibition of the thylakoid membrane by As. MDA and H_2O_2 content were increased by As, but anthocyanin content was enhanced with bacterial combination into *Azolla* and their stressors markers that were observed lower than other treatment groups. It shows that *Azolla* created tolerance against As by increasing anthocyanin content. In anthocyanin content, an increase was observed in *Azolla* in As treatment, while a decrease was observed in bacterial combination. As accumulation decreased in the *Azolla* with bacteria combination (Agnihotri, Maitra, and Mitra, 2022). *Azolla* is used in rice fields due to its fertilizer and phytoremediation potential. In a study with a combination of *Azolla* and bacteria, it was observed that As accumulation of rice decreased. Shoot elongation of rice in the combination group was higher than in the control (rice) group. The positive effect on rice physiology compared to other treatment groups may be due to IAA produced by bacteria and fixed nitrogen provided by *Azolla* (Agnihotri et al., 2022).

Cr

Cr toxic effects depend on their transformation feature within the plant's system. Cr exhibits toxic effects on plants such as yield reduction, inhibition of enzymatic activity, and mutagenesis (Shanker et al., 2005). When investigating the tolerance of *S. polyrhiza* and *A. pinnata* to the same HMs, both plants showed the highest accumulation of Ni and the lowest accumulation of Cr. Despite the low accumulation of Cr, its toxic effect on plants is highest after Cd at increasing concentrations (Gaur, Noraho, and Chauhan, 1994). Cr ions accumulated in *Azolla* proportionally to the increasing concentrations (Bennicelli et al., 2004). As the concentrations of Cr applied to different *Azolla* species increased, the fresh weight of the plants decreased inversely (Arora, Saxena, and Sharma, 2005; Bianchi et al., 2020). In the application of Cr(III), there was no significant change in the biomass content of *A. pinnata* compared to the control group at low concentrations of Cr(III), but a decrease was observed at increasing concentrations, as well as in the application of Cr(IV) groups. Cr(VI) accumulation was rapid at low concentrations at the beginning of the exposure period, while it remained stable after a certain period at high concentrations (Rai, 2009c). However, in another study examining the effects of different concentrations of HMs applied to *A. pinnata*, it was observed that Cr had less toxic effects than all other HMs, including high concentrations of Cr (Sarkar and Jana, 1986). The remediation of wastewater containing various HMs has been investigated using *Azolla* (Bharti and Banerjee, 2012). Levels of many HMs, as well as essantial elements from various wastewater sources were detected to be above the levels determined by the EPA. Upon application of wastewater to *A. pinnata*, there was a significant (~93%) decrease in Cr content of the test solution by the end of the exposure period (Vaseem and Banerjee, 2012). In a study conducted with fly ash, it was found that terrestrial plants accumulate more HMs compared to macrophytes. However, among the macrophytes, *A. pinnata* was identified as having hyperaccumulator potential for Cr HM (Kumari, Lal, and Rai, 2016).

Hg

Hg, which is not essential for plants, accumulates predominantly in the roots. Hg most likely enters into root cells via Fe, Cu, or Zn transporters or channels. It has a toxic effect on the cell, promoting the production of ROS and reducing the photosynthetic efficiency (Chen and Yang, 2012). When the effect of different HM combinations on Hill activity was examined, the lowest Hill activity compared to the control group was detected in Hg-containing combinations. Hg may have a negative effect on the photochemical reaction of PSII. This negative effect may be realized by damaging the chloroplast membrane (Sarkar and Jana, 1987). Hg accumulation in *Azolla* is rapid in the early part of the exposure period and picks up with increasing Hg concentration (Mishra, Nanda, and Misra, 1987; Bennicelli et al., 2004; Rai, 2008; Mishra, Tripathi, and Kim, 2009). In *Azolla*, as Hg concentration increased, the content of chlorophyll-a, protein, along with DNA and RNA contents, all decreased due to the toxic effect of Hg. At the same time, NO^{-3} and PO_4^{-3} uptake into the plant body decreased compared to the control group (Rai and Tripathi., 2009). The toxic effect of Hg was also toxic to cyanobacteria, the symbiont of *Azolla*.

Cu

High concentrations of Cu, an essential element for plants, demonstrate a toxic effect on plant growth, leaf chlorosis, and cytotoxicity. Excessive accumulation of Cu in the plant increases the production of ROS, a stress factor. Accordingly, the plant reduces Cu toxicity by increasing antioxidant activity (Kumar et al., 2021). When the effects of different concentrations of HMs applied to *A. pinnata* on the plant test solution were examined, it was observed that the toxic effect of Cu on the plant at increasing concentrations was higher than As, Pb and Cr (Sarkar and Jana, 1986). Like many HMs, metallothionein is involved in the transport of Cu in the plant. The expression of *AzMT2* and *PCS2* genes compared to the other groups, including the control, increased in the Cu group depending on concentration and the exposure period (Schor-Fumbarov et al., 2005; Talebi, Tabatabaei and Akbarzadeh, 2019). In a study with a different *Azolla* species, it was found that Cu accumulation increased with increasing concentrations, while the biomass decreased. *A. pinnata* was found to be the least Cu tolerant species (Akhtar et al., 2022). Despite this, the highest BCF value was reached with *A. pinnata*. With increasing concentrations of Cu and the exposure time, cell permeability increases together with the percentage of electrolyte leakage. With this cell damage, high concentrations of Cu treatment decreased the potent photochemical yield. High concentrations of Cu may have caused structural changes in the photosynthetic apparatus of *Azolla* (Sánchez-Viveros et al., 2010). The phenolic content and the maximum photochemical efficiency of PSII decreased with increasing concentrations of Cu. The higher the concentrations of Cu applied to *A. filiculoides*, the higher its accumulation in the plant body. Accordingly, high concentrations caused toxic effects and decreased the Fv/Fm value (Valderrama et al., 2012). As shown in the study for various HMs (Bharti and Banerjee, 2012), levels of many HMs, as well as essential elements from various wastewater sources were detected to be above the levels determined by the EPA. Upon application of wastewater to *A. pinnata*, there was a significant (~97%) decrease in the Cu content of the test solution by the end of the exposure period (Vaseem and Banerjee, 2012). More Cu accumulation occurred compared to *L. minor* (Vaseem and Banerjee, 2012). In various combination treatments of Cu, Si, and As, there was no significant difference in Cu accumulation of *A. caroliniana* also in the case of Cu+Si combination. In the presence of As, Cu accumulation decreased significantly. However Cu treatment did not affect the biomass content compared to the control group on *Azolla*. Since Cu application did not have a toxic effect on *A. caroliniana*, total anthocyanidins could not be detected and plant frond color change did not occur: Accordingly, there was no significant difference in chlorophyll a+b content compared to the control group. Cu treatment decreased biomass, RGR, and chlorophyll a+b content and increased total anthocyanidin in As combinations (Rofkar, Dwyer, and Bobak, 2014; Abdulwahab Al-Baldawi et al., 2022).

Zn

High concentrations of the essential element Zn can induce toxic effects in plants such as shoot shortening, leaf curling, bud death, and chlorosis (Rout and Das, 2009). The accumulation of Zn in plants has been found to be higher in *A. filiculoides* compared to Cr, Cu, and Ni groups, depending on the pH-level and the concentration of the HM (Sela, Garty, and Tel-Or, 1989). The reason behind this is the higher mobility of Zn and Cd compared to other metals used in applications. The expression of the *AzMT2* gene has been observed to increase with rising concentrations of Zn and then stabilize over time. Additionally, the expression of the *AzMT2* gene is higher in the presence of the essential element Cu, and non-essential HMs such as Cd and Ni (Schor-Fumbarov et al., 2005). The accumulation of Zn in *A. filiculoides*, *A. pinnata*, and Anzali increase with higher concentrations of Zn applied, leading to an increase in the expression of *MT2* and *PCS1* genes over the exposure period (Talebi, Tabatabaei, and Akbarzadeh, 2019). As shown in the study for various HMs (Bharti and Banerjee, 2012), levels of many HMs, as well as essential elements from various wastewater sources were detected to be above the levels determined by the EPA. Upon application of wastewater to *A. pinnata*, there was a significant decrease in the Zn content of the test solution by the first day of the exposure period (Vaseem and Banerjee, 2012).

Conclusion

Many studies have shown that *Azolla* accumulates organic and inorganic pollutants and shows tolerance to pollutants at certain concentrations (Kösesakal and Seyhan, 2023a; Kösesakal and Seyhan, 2023b; Kösesakal et al., 2016; Masoudian, Salehi-Lisar and Norastehnia, 2020; Zazouli et al., 2014; Pandey, 2012). Previous studies have revealed that *Azolla* has phytoremediation potential (Sood et al., 2012). The main feature that differentiates hyperaccumulators from non-hyperaccumulators is the ability to efficiently transport HMs from root-to-shoot. It has been acknowledged that hyperaccumulators show hypertolerance by overexpressing the transport proteins and phytochelate molecules, as well as exhibiting antioxidant activity. The research indicated that *Azolla* can accumulate HMs (Table 4) in the shoot regions such as bundle cells, bundle xylem cells, and outer cell wall of the epidermis (Sela, Garty, and Tel-Or, 1989; Sela et al., 1990; Benaroya et al., 2004). Sela et al., (1988) showed that Cd was accumulated significantly in the *Azolla*'s bundle xylem cells. As a result, the transportation of HMs to the shoot regions of *Azolla* occurs via the symplastic way. It is known that HMs are translocated by transport membrane proteins within the plant. Previous studies declared that non-essential HMs, including Cd, Pb, and Hg, were translocated from cell to cell or within a cell via some transport proteins such as the ZIP-family and the ATPase. In high concentrations of Pb accumulation in *Azolla*, overexpression of ATPase was detected.

Plants have developed tolerance strategies againts HM -induced stress. One of the tolerance strategies is phytochelatins, which are bind to HMs and translocate them to some regions of the cell such as the vacuolar and the cell wall. Overexpression of certain genes responsible for phytochelatin biosynthesis (PCS and MT2) has been observed in *Azolla* under increasing concentrations of HMs. Additionally, the content of molecules exhibiting antioxidant properties has increased (Dai et al., 2006; Dai, Dong, and Ma, 2012a). Dai, Dong, and Ma (2012b), detected an increase in the expression of *CHS* and *DFR* genes, which are responsible for anthocyanin biosynthesis, due to Cd accumulation in *Azolla*. However, in contrast to the tolerance mechanism, HMs also exhibit toxic effects on the *Azolla* physiology. These effects include a decrease in biomass content and the production of ROS that can harm the cell membrane structure, leading to the disruption of photosynthetic activity by damaging the electron transport system in chloroplast. In addition to knowledge of the reaction of *Azolla* against the toxic effects of HMs, Sarkar and Roy (2024) reported that metabolome profile, including stress-responsive metabolites, were brought to light under bisphenol A in *A. filiculoides*.

Table 4. Phytoremediation studies on *Azolla* species exposure to HMs

Species	**HMs**	**Initial concentration of HMs**	**Duration of experiment (days)**	**HMs accumulated (dry weight basis)**	**References**
A. caroliniana	Hg (II)	0.1 mg dm^{-3}	12 days	70.8 mg dm^{-3}	(Bennicelli et al., 2004)
		0.5 mg dm^{-3}		306 mg dm^{-3}	
		1 mg dm^{-3}		578 mg dm^{-3}	
	Cr (III)	0.1 mg dm^{-3}		83.5 mg dm^{-3}	
		0.5 mg dm^{-3}		412 mg dm^{-3}	
		1 mg dm^{-3}		964 mg dm^{-3}	
	Cr (VI)	0.1 mg dm^{-3}		91.1 mg dm^{-3}	
		0.5 mg dm^{-3}		157 mg dm^{-3}	
		1 mg dm^{-3}		356 mg dm^{-3}	
A. pinnata	Hg	0.1 mg L^{-1}	6 days	71 mg kg^{-1}	(Rai and Tripathi, 2009)
		0.5 mg L^{-1}		308 mg kg^{-1}	
		1.0 mg L^{-1}		580 mg kg^{-1}	
		3.0 mg L^{-1}		940 mg kg^{-1}	
A. filiculoides	Pb	5 mg L^{-1}	28 days	636 ± 224 mg kg^{-1}	(Veselý, Tlustoš and Száková, 2011)
		10 mg L^{-1}		475 ± 121 mg kg^{-1}	
A. pinnata	Hg (II)	0.5 mg L^{-1}	13 days	310 mg kg^{-1}	(Rai, 2008)
		1.0 mg L^{-1}		580 mg kg^{-1}	
		3.0 mg L^{-1}		667 mg kg^{-1}	
	Cd (II)	0.5 mg L^{-1}		408.2 mg kg^{-1}	
		1.0 mg L^{-1}		650 mg kg^{-1}	
		3.0 mg L^{-1}		740 mg kg^{-1}	
A. pinnata	Cr (III) and Cr (IV)	0.5 mg L^{-1}	13 days	415-1095 mg kg^{-1}	(Rai, 2009c)
		1.0 mg L^{-1}			
		3.0 mg L^{-1}			

Species	HMs	Initial concentration of HMs	Duration of experiment (days)	HMs accumulated (dry weight basis)	References
A. pinnata	Hg	10 μg L^{-1}	7 days	0.58 ± 0.01 mg g^{-1}	(Mishra, Tripathi and Kim, 2009)
			14 days	0.69 ± 0.01 mg g^{-1}	
			21 days	0.75 ± 0.01 mg g^{-1}	
			28 days	0.77 ± 0.01 mg g^{-1}	
A. caroliniana	Pb (II)	0.1 mg dm^{-3}	12 days	53-416 mg kg^{-1}	(Stepniewska et al., 2005)
		0.5 mg dm^{-3}			
		1 mg dm^{-3}			
	Cd (II)	0.1 mg dm^{-3}		23-259 mg kg^{-1}	
		0.5 mg dm^{-3}			
		1 mg dm^{-3}			
A. imbricata	Cd	0.01 mg L^{-1}	1 day	1.079 ± 0.126 μg g^{-1}	(Dai et al., 2006)
		0.05 mg L^{-1}		6.712 ± 0.178 μg g^{-1}	
		0.1 mg L^{-1}		14.41 ± 1.542 μg g^{-1}	
		0.5 mg L^{-1}		52.23 ± 2.274 μg g^{-1}	
		0.01 mg L^{-1}	3 days	1.760d ± 0.194 μg g^{-1}	
		0.05 mg L^{-1}		14.309c ± 2.198 μg g^{-1}	
		0.1 mg L^{-1}		41.780b ± 5.743 μg g^{-1}	
		0.5 mg L^{-1}		139.369a ± 4.649 μg g^{-1}	
		0.01 mg L^{-1}	5 days	2.215c ± 0.194 μg g^{-1}	
		0.05 mg L^{-1}		18.138c ± 2.387 μg g^{-1}	
		0.1 mg L^{-1}		46.232b ± 2.364 μg g^{-1}	
		0.5 mg L^{-1}		153.579a ± 22.932 μg g^{-1}	
		0.01 mg L^{-1}	7 days	2.357c ± 0.841 μg g^{-1}	
		0.05 mg L^{-1}		19.418c ± 2.733 μg g^{-1}	
		0.1 mg L^{-1}		54.590b ± 5.913 μg g^{-1}	
		0.5 mg L^{-1}		165.782a ± 20.465 μg g^{-1}	

Table 4. (Continued)

Species	HMs	Initial concentration of HMs	Duration of experiment (days)	HMs accumulated (dry weight basis)	References
A. imbricata	Cd	0.01 mg L^{-1}	9 days	2.762c ± 0.579 μg g^{-1}	(Dai et al., 2006)
		0.05 mg L^{-1}		20.369c ± 0.611 μg g^{-1}	
		0.1 mg L^{-1}		66.729b ± 4.781 μg g^{-1}	
		0.5 mg L^{-1}		183.738a ± 25.042 μg g^{-1}	
A. microphylla	Cr	1 μg ml^{-1}	7 days	4617.7±235.2 μg g^{-1}	(Arora et al., 2006)
		5 μg ml^{-1}		6156.7 ± 379.1 μg g^{-1}	
		10 μg ml^{-1}		9213.5 ± 1043 μg g^{-1}	
		15 μg ml^{-1}		12874.3 ± 154.9 μg g^{-1}	
		20 μg ml^{-1}		14931.7 ± 2006.2 μg g^{-1}	
A. pinnata		1 μg ml^{-1}		528.1 ± 24.1 μg g^{-1}	
		5 μg ml^{-1}		1554.7 ± 538.8 μg g^{-1}	
		10 μg ml^{-1}		2434.3 ± 1663.1 μg g^{-1}	
		15 μg ml^{-1}		5507.7 ± 320.5 μg g^{-1}	
		20 μg ml^{-1}		9125.3 ± 901.3 μg g^{-1}	
A. filiculoides		1 μg ml^{-1}		2977.9 ± 26.1 μg g^{-1}	
		5 μg ml^{-1}		4122 ± 476.3 μg g^{-1}	
		10 μg ml^{-1}		6567 ± 537 μg g^{-1}	
		15 μg ml^{-1}		9994.6 ± 899.1 μg g^{-1}	
		20 μg ml^{-1}		12383.6 ± 2025 μg g^{-1}	
A. filiculoides	Cd	15 ppm	4 days	10441 ± 740 ppm	(Sela et al., 1990)
	Cr			1904 + 126 ppm	
	Cu			9224 + 109 ppm	
	Ni			8814 + 15 ppm	
	Zn			6408 + 109 ppm	

Species	HMs	Initial concentration of HMs	Duration of experiment (days)	HMs accumulated (dry weight basis)	References
A. caroliniana	Si	2 µM Cu + 1.8 µM Si	14 days	57.5 ± 4.8 Cu mg kg^{-1}; 3124 ± 147.0 Si mg kg^{-1}	(Rofkar, Dwyer and Bobak, 2014)
	Cu	78 µM Cu		328 ± 43.2 Cu mg kg^{-1}; 107 ± 8.3 Si mg kg^{-1}	
	Cu+Si	78 µM Cu + 1.8 µM Si		332 ± 5.7 Cu mg kg^{-1}; 1968 ± 132.8 Si mg kg^{-1}	
	As	20 µM As + 2 µM Cu		76.8 ± 12.4 As mg kg^{-1}; 52.8 ± 3.6 Cu mg kg^{-1}; 77.4 ± 1.5 Si mg kg^{-1}	
	As+Si	20 µM As + 2 µM Cu + 1.8 µM Si		50.8 ± 7.6 As mg kg_{-1}; 58.1 ± 1.6 Cu mg kg^{-1}; 3190 ± 376.2 Si mg kg^{-1}	
	As+Cu	20 µM As + 78 µM Cu		53.8 ± 8.8 As mg kg^{-1}; 229 ± 29.4 Cu mg kg^{-1}; 115 ± 8.9 Si mg kg^{-1}	
	As+Cu+Si	20 µM As + 78 µM Cu + 1.8 µM Si		33.7 ± 1.1 As mg kg^{-1}; 258 ± 6.2 Cu mg kg^{-1}; 3006 ± 343.7 Si mg kg^{-1}	
Anzali	Zn	500 µM	3 days	3596.79 mg kg^{-1}	(Talebi et al., 2019)
	Cu			3542.2 mg kg^{-1}	
	Cd			2309.4 mg kg^{-1}	
	Ni			937.5 mg kg^{-1}	
A. filiculoides	Zn			2563.095 mg kg^{-1}	
	Cu			2921.5 mg kg^{-1}	
	Cd			2747.1 mg kg^{-1}	
	Ni			1213.95 mg kg^{-1}	
A. pinnata	Zn			652.845 mg kg^{-1}	
	Cu			4805.2 mg kg^{-1}	
	Cd			4673.8 mg kg^{-1}	
	Ni			1030.05 mg kg^{-1}	

Table 4. (Continued).

Species	HMs	Initial concentration of HMs	Duration of experiment (days)	HMs accumulated (dry weight basis)	References
A. filiculoides (roots)	Cd	5 mg L^{-1}	5 days	2299.43 ± 31.20 μg g^{-1}	(Spanò et al., 2019)
		10 mg L^{-1}		4203.95 ± 67.26 μg g^{-1}	
A. filiculoides (frond)		5 mg L^{-1}		74.51 ± 6.52 μg g^{-1}	
		10 mg L^{-1}		649.95 ± 13.09 μg g^{-1}	
A. pinnata	Hg	0.1 mg L^{-1}	6 days	71 mg kg^{-1}	(Rai and Tripathi., 2009a)
		0.5 mg L^{-1}		308 mg kg^{-1}	
		1 mg L^{-1}		580 mg kg^{-1}	
		3 mg L^{-1}		940 mg kg^{-1}	
A. pinnata	Cd	0.05-0.5 mg L^{-1}	1 day	6-10 μg g^{-1}	(Prasad and Singh, 2011)
			4 days	10-15 μg g^{-1}	
		2 mg L^{-1}	1 day	18 μg g^{-1}	
			4 days	22 μg g^{-1}	
A. cristata	Se(IV), Se(VI) and Zn	0.5 mg L^{-1} Se (IV) + 0 mg L^{-1} Zn	5 days	29.2 ± 0.1 μg pot^{-1}	(Li et al., 2020)
		0.5 mg L^{-1} Se (IV) + 0.5 mg L^{-1} Zn		23.4 ± 0.3 μg pot^{-1}	
		0.5 mg L^{-1} Se (IV) + 1 mg L^{-1} Zn		25.9 ± 0.2 μg pot^{-1}	
		0.5 mg L^{-1} Se (IV) + 2.5 mg L^{-1} Zn		25.5 ± 0.8 μg pot^{-1}	
		0.5 mg L^{-1} Se (IV) + 5.0 mg L^{-1} Zn		23.8 ± 0.8 μg pot^{-1}	
		2.5 mg L^{-1} Se (IV) + 0 mg L^{-1} Zn		88.4 ± 3.3 μg pot^{-1}	
		2.5 mg L^{-1} Se (IV) + 0.5 mg L^{-1} Zn		74.8 ± 2.3 μg pot^{-1}	
		2.5 mg L^{-1} Se (IV) + 1 mg L^{-1} Zn		74.2 ± 1.2 μg pot^{-1}	
		2.5 mg L^{-1} Se (IV) + 2.5 mg L^{-1} Zn		65.9 ± 3.7 μg pot^{-1}	
		2.5 mg L^{-1} Se (IV) + 5.0 mg L^{-1} Zn		65.6 ± 1.8 μg pot^{-1}	
		5.0 mg L^{-1} Se (IV) + 0 mg L^{-1} Zn		90.5 ± 1.2 μg pot^{-1}	
		5.0 mg L^{-1} Se (IV) + 0.5 mg L^{-1} Zn		86.6 ± 5.8 μg pot^{-1}	
		5.0 mg L^{-1} Se (IV) + 1 mg L^{-1} Zn		78.4 ± 4.3 μg pot^{-1}	
		5.0 mg L^{-1} Se (IV) + 2.5 mg L^{-1} Zn		79.9 ± 3.3 μg pot^{-1}	
		5.0 mg L^{-1} Se (IV) + 5.0 mg $^{L-1}$ Zn		80.7 ± 2.9 μg pot^{-1}	

Species	HMs	Initial concentration of HMs	Duration of experiment (days)	HMs accumulated (dry weight basis)	References
A. cristata	Se(IV), Se(VI) and Zn	0.5 mg L^{-1} Se (VI) + 0 mg L^{-1} Zn	5 days	2.2 ± 0.1 µg pot^{-1}	(Li et al., 2020)
		0.5 mg L^{-1} Se (VI) + 0.5 mg L^{-1} Zn		1.6 ± 0.1 µg pot^{-1}	
		0.5 mg L^{-1} Se (VI) + 1 mg L^{-1} Zn		1.4 ± 0.1 µg pot^{-1}	
		0.5 mg L^{-1} Se (VI) + 2.5 mg L^{-1} Zn		1.4 ± 0.2 µg pot^{-1}	
		0.5 mg L^{-1} Se (VI) + 5.0 mg L^{-1} Zn		1.5 ± 0.1 µg pot^{-1}	
		2.5 mg L^{-1} Se (VI) + 0 mg L^{-1} Zn		11.2 ± 1.7 µg pot^{-1}	
		2.5 mg L^{-1} Se (VI) + 0.5 mg L^{-1} Zn		12.4 ± 0.1 µg pot^{-1}	
		2.5 mg L^{-1} Se (VI) + 1 mg L^{-1} Zn		9.4 ± 0.9b µg pot^{-1}	
		2.5 mg L^{-1} Se (VI) + 2.5 mg L^{-1} Zn		7.9 ± 0.6 µg pot^{-1}	
		2.5 mg L^{-1} Se (VI) + 5.0 mg L^{-1} Zn		8.9 ± 1.34 µg pot^{-1}	
		5.0 mg L^{-1} Se (VI) + 0 mg L^{-1} Zn		17.3 ± 1.3 µg pot^{-1}	
		5.0 mg L^{-1} Se (VI) + 0.5 mg L^{-1} Zn		15.1 ± 0.5 µg pot^{-1}	
		5.0 mg L^{-1} Se (VI) + 1 mg L^{-1} Zn		15.3 ± 1.4 µg pot^{-1}	
		5.0 mg L-^{1} Se (VI) + 2.5 mg L^{-1} Zn		11.6 ± 0.8 µg pot^{-1}	
		5.0 mg L^{-1} Se (VI) + 5.0 mg L^{-1} Zn		11.2 ± 0.1 µg pot^{-1}	
A. caroliniana	As	50 µM	10 days	284 mg kg^{-1}	(Zhang et al., 2008)
A. filiculoides				54 mg kg^{-1}	

The toxic effects of HMs on *Azolla* and their tolerance strategies showed great variations in previous studies, which suggest that it depends on the *Azolla* species, the concentration of HMs, and the exposure period, as well as the cultivation conditions. Although HMs in higher concentrations showed toxic effects of biomass content and photosynthetic activity, *Azolla* retained the accumulation of HMs with an increasing exposure period. Moreover, *Azolla* can tolerate certain concentrations of HMs. However, at fatal levels, the accumulation rate may slow down or stop due to the serious damage to their physiological mechanisms. Biosorption studies have shown that non-living *Azolla* can be modified to adsorb HMs through various interactions to increase their adsorption potential. The accumulation capacity and adsorption potential of *Azolla*, as well as its modified non-living forms can be evaluated in future studies depending on the concentration of HMs and the density of *Azolla* used. With the conduct of studies like those, it is believed that elucidating the tolerance mechanisms under HMs stress could facilitate the utilization of *Azolla* in remediation studies and promote its use as a biofertilizer.

Acknowledgments

This study was funded by Scientific Research Projects Coordination Unit of Istanbul University Project number: FBA-2023-39420. The authors wish to thank Cemre Şenesen, for grammatical revision of this chapter.

Disclaimer

None.

References

Agnihotri, P., Maitra, M. and Mitra, A. K. 2022. "Effect of an As (V)-Resistant Plant Growth Promoting Bacterial Consortium on Growth, Antioxidant Content and Arsenic Accumulation in *Azolla microphylla* Kaulf. Under As (V) Stress." *Russian Journal of Plant Physiology* 69(1):15.

Agnihotri, P., Sikdar, S., Maitra, M., Choudhury, S. S. and Mitra, A. K., 2022. "Effect of combination of *Azolla microphylla* and As (V)-resistant bacterial consortium on growth, oxidative stress and arsenic accumulation in rice plant under As (V) stress." *Vegetos*, 35(3):796-802.

Akhtar, M. S., Aslam, S., Ditta, A., Albalawi, B. F. A., Oki, Y. and Nakashima, Y., 2022. "Interspecific Variability in Growth Characteristics and Phytoremediation of Cu by Free-Floating *Azolla* Macrophytes." *Sustainability*, 15(1):497.

Akhter, M. F., Christopher R. Omelon, Robert A. Gordon, Desmond Moser, and Sheila M. Macfie. 2014. "Localization and chemical speciation of cadmium in the roots of barley and lettuce." *Environmental and Experimental Botany* 100:10-19.

Akinbile, C. O., Ogunrinde, T. A., Che bt Man, H. and Aziz, H. A., 2016. "Phytoremediation of Domestic Wastewaters in Free Water Surface Constructed Wetlands Using *Azolla pinnata*," *International Journal of Phytoremediation*, 18(1):54-61.

Al-Baldawi, I. A., Yasin, S. R., Jasim, S. S., Abdullah, S. R. S. and Almansoory, A. F., 2022. "Removal of copper by *Azolla filiculoides* and *Lemna minor*: phytoremediation potential, adsorption kinetics and isotherms." *Heliyon* 8(11).

Ali, S., Abbas, Z., Rizwan, M., Zaheer, I. E., Yavaş, İ., Ünay, A., Abdel-Daim, M. M., Bin-Jumah, M., Hasanuzzaman, M. and Kalderis, D., 2020. "Application of floating aquatic plants in phytoremediation of heavy metals polluted water: A review." *Sustainability*, 12(5):1927.

Amare, E., Kebede, F., Berihu, T. and Mulat, W., 2018. "Field-based investigation on phytoremediation potentials of *Lemna minor* and *Azolla filiculoides* in tropical, semiarid regions: Case of Ethiopia." *International journal of phytoremediation*, 20(10):965-972.

Amoatey, P. and Baawain, M. S. 2019. "Effects of pollution on freshwater aquatic organisms." *Water Environment Research.* 91(10):1272-1287.

Appenroth, K. J., 2010. "What are "heavy metals" in plant sciences?." *Acta Physiologiae Plantarum*. 32:615-619.

Arora, A., Saxena, S., and Sharma, D. K., 2006. "Tolerance and phytoaccumulation of Chromium by three *Azolla* species." *World Journal of Microbiology and Biotechnology* 22:97-100.

Ayala, A., Muñoz, M. F. and Argüelles, S. 2014. "Lipid peroxidation: production, metabolism, and signaling mechanisms of malondialdehyde and 4-hydroxy-2-nonenal." *Oxidative medicine and cellular longevity*, 2014.

Ayala, A., Muñoz, M. F. and Argüelles, S., 2014. "Lipid peroxidation: production, metabolism, and signaling mechanisms of malondialdehyde and 4-hydroxy-2-nonenal." *Oxidative medicine and cellular longevity*, 2014.

Bali, A. S. and Sidhu, G. P. S., 2021. "Arsenic acquisition, toxicity and tolerance in plants-From physiology to remediation: A review." *Chemosphere*, 283:131050.

Banach, A. M., Kuźniar, A., Grządziel, J. and Wolińska, A., 2020. "*Azolla filiculoides* L. as a source of metal-tolerant microorganisms." *PLoS One*, 15(5):e0232699.

Benaroya, R. O., Tzin, V., Tel-Or, E. and Zamski, E., 2004. "Lead accumulation in the aquatic fern *Azolla filiculoides*." *Plant physiology and Biochemistry*, 42(7-8):639-645.

Bennicelli, R., Stępniewska, Z., Banach, A., Szajnocha, K. and Ostrowski, J., 2004. "The ability of *Azolla caroliniana* to remove heavy metals (Hg (II), Cr (III), Cr (VI)) from municipal waste water." *Chemosphere*, 55(1):141-146.

Bhaduri, A. M. and Fulekar, M. H., 2012. "Antioxidant enzyme responses of plants to heavy metal stress." *Reviews in Environmental Science and Bio/Technology*, 11:55-69.

Bharti, S. and Banerjee, T. K., 2012. "Phytoremediation of the coalmine effluent." *Ecotoxicology and environmental safety*, 81:36-42.

Bianchi, E., Biancalani, A., Berardi, C., Antal, A., Fibbi, D., Coppi, A., Lastrucci, L., Bussotti, N., Colzi, I., Renai, L. and Scordo, C., 2020. "Improving the efficiency of wastewater treatment plants: Bio-removal of heavy-metals and pharmaceuticals by *Azolla filiculoides* and *Lemna minuta*." *Science of the Total Environment*, 746:141219.

Breckle, S. W. 1991. "Growth Under Stress: Heavy Metals" In *Plant Roots*: *The Hidden Half* edited Waisel, Y., Eshel, A., and Kafkafi, U, 351-373, New York: Dekker

Breckle, S. W. and Kahle, H. 1992. "Effects of toxic heavy metals (Cd, Pb) on growth and mineral nutrition of beech (*Fagus sylvatica* L.)." *Vegetatio* 101:43-53.

Burzyński, M., and Kolano, E. 2003. "In vivo and in vitro effects of copper and cadmium on the plasma membrane H+-ATPase from cucumber (*Cucumis sativus* L.) and maize (*Zea mays* L.) roots." *Acta Physiol Plant*, 25:39–45.

Carrapiço, F., 2010. "*Azolla* as a superorganism. Its implication in symbiotic studies." *Symbioses and Stress: Joint Ventures in Biology*, 225-241.

Catalá, A. 2009. "Lipid peroxidation of membrane phospholipids generates hydroxy-alkenals and oxidized phospholipids active in physiological and/or pathological conditions." *Chemistry and physics of lipids*, 157(1):1-11.

Chappell, J. 1997. "Phytoremediation of TCE using Populus," *Status report prepared for the U.S. EPA Technology Innovation Office under a National Network of Environmental Management Studies Fellowship* Compiled June-August 1997, http://www.cluin.org/download/studentpapers/phytotce.pdf

Chaudhary, K., Agarwal, S. and Khan, S. 2018. "Role of phytochelatins (PCs), metallothioneins (MTs), and heavy metal ATPase (HMA) genes in heavy metal tolerance." *Mycoremediation and Environmental Sustainability*. 2:39-60.

Chen, J. and Yang, Z. M., 2012. "Mercury toxicity, molecular response and tolerance in higher plants." *Biometals*, 25:847-857.

Citterio, S., Santagostino, A., Fumagalli, P., Prato, N., Ranalli, P. and Sgorbati, S., 2003. Heavy metal tolerance and accumulation of Cd, Cr and Ni by Cannabis sativa L. Plant and soil, 256, pp.243-252. 2003. "Heavy metal tolerance and accumulation of Cd, Cr and Ni by *Cannabis sativa* L." *Plant and soil*, 256:243-252.

Clabeaux, B. L., Navarro, D. A., Aga, D. S. and Bisson, M. A., 2011. "Cd tolerance and accumulation in the aquatic macrophyte, *Chara australis*: potential use for charophytes in phytoremediation." *Environmental science & technology*, 45(12):5332-5338.

Clemens, S. and Ma, J. F., 2016. "Toxic heavy metal and metalloid accumulation in crop plants and foods." *Annual review of plant biology*. 67:489-512.

Dai, L. P., Dong, X. J. and Ma, H. H., 2012a. "Antioxidative and chelating properties of anthocyanins in *Azolla imbricata* induced by cadmium." *Polish Journal of Environmental Studies*, 21(4).

Dai, L. P., Dong, X. J. and Ma, H. H., 2012b. "Molecular mechanism for cadmium-induced anthocyanin accumulation in *Azolla imbricata*." *Chemosphere*, 87(4):319-325.

Dai, L. P., Xiong, Z. T., Huang, Y. and Li, M. J., 2006. "Cadmium-induced changes in pigments, total phenolics, and phenylalanine ammonia-lyase activity in fronds of *Azolla imbricata*." *Environmental Toxicology: An International Journal*, 21(5):505-512.

Das, K. and Roychoudhury, A. 2014 "Reactive oxygen species (ROS) and response of antioxidants as ROS-scavengers during environmental stress in plants." *Frontiers in environmental science* 2:53.

de Paiva Magalhães, D., da Costa Marques, M. R., Baptista, D. F. and Buss, D. F., 2015. "Metal bioavailability and toxicity in freshwaters."*Environmental Chemistry Letters."* 13(1):69-87.

Devi, S. R., Prasad, M. N. V. 1999 "Membrane Lipid Alterations in Heavy Metal Exposed Plants." In *Heavy Metal Stress in Plants,* edited by M. N. V. Prasad and J. Hagemeyer, 99-116 Springer, Berlin, Heidelberg.

Fahad, S., Rehman, A., Shahzad, B., Tanveer, M., Saud, S., Kamran, M., Ihtisham, M., Khan, S. U., Turan, V. and ur Rahman, M. H., 2019. "Rice responses and tolerance to metal/metalloid toxicity." In *Advances in rice research for abiotic stress tolerance* edited by Mirza Hasanuzzaman, Masayuki Fujita, Kamrun Nahar and Jiban Krishna Biswas 299-312. Woodhead publishing.

Farooq, M. A., Islam, F., Ali, B., Najeeb, U., Mao, B., Gill, R. A., Yan, G., Siddique, K. H. and Zhou, W., 2016. "Arsenic toxicity in plants: cellular and molecular mechanisms of its transport and metabolism." *Environmental and Experimental Botany*, 132:42-52.

Fodor, E., Szabó-Nagy, A. and Erdei, L. 1995. "The effects of cadmium on the fluidity and H+-ATPase activity of plasma membrane from sunflower and wheat roots." *Journal of plant physiology,* 147(1):87-92.

Forni, C., Braglia, R., Harren, F. J. M. and Cristescu, S. M., 2012. "Stress Responses of Duckweed (*Lemna mino*r L.) and Water Velvet (*Azolla filiculoides* Lam.) to Anionic Surfactant Sodium-Dodecyl-Sulphate (Sds)," *Aquatic Toxicology*, 110(111):107-113.

Gaur, J. P., Noraho, N. and Chauhan, Y. S., 1994. "Relationship between heavy metal accumulation and toxicity in *Spirodela polyrhiza* (L.) Schleid. and *Azolla pinnata* R. Br." *Aquatic Botany*" 49(2-3):183-192.

Gekeler, W., Grill, E., Winnacker, E. L. and Zenk, M. H., 1989. "Survey of the plant kingdom for the ability to bind heavy metals through phytochelatins." *Zeitschrift für Naturforschung C.* 44(5-6):361-369.

Giller, K. E., Witter, E. and Mcgrath, S. P. 1998. "Toxicity of heavy metals to microorganisms and microbial processes in agricultural soils: a review." *Soil biology and biochemistry.* 30(10-11):1389-1414.

Guimaraes, F. P., Aguiar, R., Oliveira, J. A., Silva, J. A. A. and Karam, D., 2012. "Potential of macrophyte for removing arsenic from aqueous solution." *Planta daninha*, 30:683-696.

Gusman, G. S., Oliveira, J. A., Farnese, F. S. and Cambraia, J., 2013. "Arsenate and arsenite: the toxic effects on photosynthesis and growth of lettuce plants." *Acta Physiologiae Plantarum*, 35:1201-1209.

Hassinen, V. H., Tervahauta, A. I., Schat, H. and Kärenlampi, S. O., 2011. "Plant metallothioneins–metal chelators with ROS scavenging activity?." *Plant Biology*. 13(2):225-232.

Imran, M. A., Ch, M. N., Khan, R. M., Ali, Z. and Mahmood, T., 2013. "Toxicity of arsenic (As) on seed germination of sunflower (*Helianthus annuus* L.)." *Int. J. Phys. Sci.* 8(17):840-847.

Inouhe, M., Huang, H., Chaudhary, S. K. and Gupta, D. K., 2012. "Heavy metal bindings and their interactions with thiol peptides and other biological ligands in plant cells." In *Metal toxicity in plants: perception, signaling and remediation*, edited by Dharmendra K. Gupta and Luisa M. Sandalio, 1-21. Heidelberg: Springer Berlin

Iriel, A., Dundas, G., Cirelli, A. F. and Lagorio, M. G., 2015. "Effect of arsenic on reflectance spectra and chlorophyll fluorescence of aquatic plants." *Chemosphere* 119:697-703.

Ismael, M. A., Elyamine, A. M., Moussa, M. G., Cai, M., Zhao, X. and Hu, C., 2019. "Cadmium in plants: uptake, toxicity, and its interactions with selenium fertilizers." *Metallomics*, 11(2):255-277.

Jambunathan, N. 2010. "Determination and detection of reactive oxygen species (ROS), lipid peroxidation, and electrolyte leakage in plants." *Methods in molecular biology* (Clifton, N. J.), 639:292-8.

Janicka-Russak, M., Kabała, K., Burzyński, M. and Kłobus, G., 2008. "Response of plasma membrane H+-ATPase to heavy metal stress in *Cucumis sativus* roots. "*Journal of Experimental Botany*" 59(13):3721-3728.

Kösesakal, T. and Seyhan, M., 2023a. "Phenanthrene stress response and phytoremediation potential of free-floating fern *Azolla filiculoides* Lam" *International Journal of Phytoremediation*, 25(2):207-220.

Kösesakal, T. and Seyhan, M., 2023b. "Naphthalene stress responses of the aquatic fern *Azolla filiculoides* Lam. and evaluation of phytoremediation potential." *Polycyclic Aromatic Compounds*, 43(8):6819-6836.

Kösesakal, T., Ünal, M., Kulen, O., Memon, A. and Yüksel, B., 2016. "Phytoremediation of Petroleum Hydrocarbons by Using a Freshwater Fern Species *Azolla filiculoides* Lam.," *International Journal of Phytoremediation*, 18(5):467-476.

Krupa, Z., 1999. "Cadmium against higher plant photosynthesis-a variety of effects and where do they possibly come from?." *Zeitschrift für Naturforschung C*, 54(9-10):723-729.

Kumar, A. and Prasad, M. N. V., 2018. "Plant-lead interactions: transport, toxicity, tolerance, and detoxification mechanisms." *Ecotoxicology and environmental safety*, 166:401-418.

Kumar, V., Pandita, S., Sidhu, G. P. S., Sharma, A., Khanna, K., Kaur, P., Bali, A. S. and Setia, R., 2021. "Copper bioavailability, uptake, toxicity and tolerance in plants: A comprehensive review." *Chemosphere*, 262:127810.

Kumari, A., Lal, B. and Rai, U. N., 2016. "Assessment of native plant species for phytoremediation of heavy metals growing in the vicinity of NTPC sites, Kahalgaon, India." *International journal of phytoremediation, 18*(6):592-597.

Leão, G. A., Oliveira, J. A. D., Felipe, R. T. A. and Farnese, F. S., 2017. "Phytoremediation of arsenic-contaminated water: the role of antioxidant metabolism of *Azolla caroliniana* Willd.(Salviniales)." *Acta Botanica Brasilica*, 31:161-168.

Li, J., Lens, P. N., Otero-Gonzalez, L. and Du Laing, G., 2020. "Production of selenium- and zinc-enriched *Lemna* and *Azolla* as potential micronutrient-enriched bioproducts." *Water research*, 172:115522.

Liu, C., Guo, B., Li, H., Fu, Q., Li, N., Lin, Y. and Xu, G., 2021. "*Azolla* incorporation under flooding reduces grain cadmium accumulation by decreasing soil redox potential." *Scientific Reports*, 11(1):6325.

Manisalidis, I., Stavropoulou, E., Stavropoulos, A. and Bezirtzoglou, E., 2020. "Environmental and health impacts of air pollution: a review." *Frontiers in public health*, 8:14.

Mansoor, S., Ali, A., Kour, N., Bornhorst, J., AlHarbi, K., Rinklebe, J., Abd El Moneim, D., Ahmad, P. and Chung, Y. S., 2023. "Heavy Metal Induced Oxidative Stress Mitigation and ROS Scavenging in Plants." *Plants*, 12(16):3003.

Mashkani, S. G. and Ghazvini, P. T. M., 2009. "Biotechnological potential of *Azolla filiculoides* for biosorption of Cs and Sr: Application of micro-PIXE for measurement of biosorption." *Bioresource technology*, 100(6):1915-1921.

Masoudian, Z., Salehi-Lisar, S. Y. and Norastehnia, A., 2020. "Phytoremediation potential of *Azolla filiculoides* for sodium dodecyl benzene sulfonate (SDBS) surfactant considering some physiological responses, effects of operational parameters and biodegradation of surfactant." *Environmental Science and Pollution Research*, 27:20358-20369.

Mishra, B. B., Nanda, D. R. and Misra, B. N., 1987. "Accumulation of mercury by *Azolla* and its effect on growth." *Bull. Environ. Contam. Toxicol.;(United States)*, 39(4).

Mishra, V. K., Tripathi, B. D. and Kim, K. H., 2009. "Removal and accumulation of mercury by aquatic macrophytes from an open cast coal mine effluent." *Journal of Hazardous materials*, 172(2-3):749-754.

Mitra, A., Chatterjee, S., Datta, S., Sharma, S., Veer, V., Razafindrabe, B. H., Walther, C. and Gupta, D. K., 2014. "Mechanism of metal transporters in plants." *Heavy metal remediation: transport and accumulation in plants*, 1-28.

Naghipour, D., Ashrafi, S. D., Gholamzadeh, M., Taghavi, K. and Naimi-Joubani, M., 2018. "Phytoremediation of heavy metals (Ni, Cd, Pb) by *Azolla filiculoides* from aqueous solution: A dataset." *Data in brief*, 21:1409-1414.

Naing, A. H. and Kim, C. K. 2021. "Abiotic stress-induced anthocyanins in plants: Their role in tolerance to abiotic stresses." *Physiologia Plantarum*, 172(3):1711-1723.

Nie, J., Liu, Y., Zeng, G., Zheng, B., Tan, X., Liu, H., Xie, J., Gan, C. and Liu, W., 2016. "Cadmium accumulation and tolerance of *Macleaya cordata*: a newly potential plant for sustainable phytoremediation in Cd-contaminated soil." *Environmental Science and Pollution Research*. 23:10189-99.

Pandey, P., Singh, J., Achary, V. M. M. and Reddy, M. K., 2015. "Redox homeostasis via gene families of ascorbate-glutathione pathway." *Frontiers in Environmental Science*, 3:25.

Pandey, V. C., 2012. "Phytoremediation of heavy metals from fly ash pond by *Azolla caroliniana.*" *Ecotoxicology and Environmental Safety*, 82:8-12.

Pang, Y. L., Quek, Y. Y., Lim, S. and Shuit, S. H., 2023. "Review on phytoremediation potential of floating aquatic plants for heavy metals: a promising approach." *Sustainability*, 15(2):1290.

Parys, E., Romanowska, E., Siedlecka, M. and Poskuta, J. W., 1998. "The effect of lead on photosynthesis and respiration in detached leaves and in mesophyll protoplasts of *Pisum sativum.*" *Acta Physiologiae Plantarum*, 20:313-322.

Prasad, S. M. and Singh, A., 2011. "Metabolic responses of *Azolla pinnata* to cadmium stress: photosynthesis, antioxidative system and phytoremediation." *Chemistry and ecology*, 27(6):543-555.

Rachmadiarti, F., Fitrihidajati, H., Purnomo, T., Yuliani, Y. and Wahyuningsih, D. A., 2018. "*Azolla microphylla* and *Pistia stratiotes* as Phytoremediator of Pb (lead)." In *International Conference on Science and Technology (ICST 2018),* 92-95, Atlantis Press.

Rai, P. K. and Tripathi, B. D., 2009. "Comparative assessment of *Azolla pinnata* and *Vallisneria spiralis* in Hg removal from GB Pant Sagar of Singrauli Industrial region, India." *Environmental Monitoring and Assessment*, 148:75-84.

Rai, P. K., 2008. "Phytoremediation of Hg and Cd from industrial effluents using an aquatic free floating macrophyte *Azolla pinnata.*" *International journal of phytoremediation*, 10(5):430-439.

Rai, P. K., 2009b. "Heavy metals in water, sediments and wetland plants in an aquatic ecosystem of tropical industrial region, India." *Environmental Monitoring and Assessment*, 158:433-457.

Rai, P. K., 2009c. "Microcosm investigation on phytoremediation of Cr using *Azolla pinnata.*" *International journal of phytoremediation*, 12(1):96-104.

Ran, L., Larsson, J., Vigil-Stenman, T., Nylander, J. A., Ininbergs, K., Zheng, W. W., Lapidus, A., Lowry, S., Haselkorn, R. and Bergman, B., 2010. "Genome Erosion in a Nitrogen-Fixing Vertically Transmitted Endosymbiotic Multicellular Cyanobacterium." *PLOS ONE* 5:e11486.

Rascio, N. and Navari-Izzo, F., 2011. "Heavy metal hyperaccumulating plants: how and why do they do it? And what makes them so interesting?." *Plant science.* 180(2):169-181.

Rezania, S., Taib, S. M., Din, M. F. M., Dahalan, F. A. and Kamyab, H., 2016. "Comprehensive review on phytotechnology: Heavy metals removal by diverse aquatic plants species from wastewater." *J Hazard Mater* 318:587-599.

Roberts, A. E., Boylen, C. W. and Nierzwicki-Bauer, S. A., 2014. "Effects of lead accumulation on the *Azolla caroliniana*–Anabaena association." *Ecotoxicology and Environmental safety*, 102:100-104.

Rofkar, J. R., Dwyer, D. F. and Bobak, D. M., 2014. "Uptake and toxicity of arsenic, copper, and silicon in *Azolla caroliniana* and *Lemna minor.*" *International journal of phytoremediation*, 16(2):155-166.

Romero-Hernández, J. A., Amaya-Chávez, A., Balderas-Hernández, P., Roa-Morales, G., González-Rivas, N. and Balderas-Plata, M. Á., 2017. "Tolerance and hyperaccumulation of a mixture of heavy metals (Cu, Pb, Hg, and Zn) by four aquatic macrophytes." *International journal of phytoremediation*, 19(3):239-245.

Rout, G. R. and Das, P., 2009. "Effect of metal toxicity on plant growth and metabolism: I. Zinc." *Sustainable agriculture* 873-884.

Sánchez-Viveros, G., Ferrera-Cerrato, R. and Alarcón, A., 2011. "Short-term effects of arsenate-induced toxicity on growth, chlorophyll and carotenoid contents, and total content of phenolic compounds of *Azolla filiculoides*." *Water, Air, & Soil Pollution*, 217:455-462.

Sanchez-Viveros, G., Gonzalez-Mendoza, D., Alarcon, A. and Ferrera-Cerrato, R., 2010. "Copper effects on photosynthetic activity and membrane leakage of *Azolla filiculoides* and *A. caroliniana*." *International Journal of Agriculture and Biology*, 12(3):365-368.

Sánchez-Viveros, G., Ruvalcaba-Sil, J. L., Ferrera-Cerrato, R., Alarcón, A. and Xoconostle-Cázares, B., 2016. "Changes in elemental content in fronds of *Azolla filiculoides* due to arsenic accumulation." *Plant Biosystems-An International Journal Dealing with all Aspects of Plant Biology*, 150(6):1332-1340.

Sarkar, A. and Jana, S., 1986. "Heavy metal pollutant tolerance of *Azolla pinnata*." *Water, Air, and Soil Pollution*, 27:15-18.

Sarkar, A. and Jana, S., 1987. "Effects of combinations of heavy metals on Hill activity of *Azolla pinnata*." *Water, Air, and Soil Pollution*, 35:141-145.

Sarkar, A. and Roy, S., 2024. "Metabolome profile variation in *Azolla filiculoides* exposed to Bisphenol A assists in the identification of stress-responsive metabolites." *Aquatic Toxicology*, 266:106792.

Schat, H., Llugany, M., Vooijs, R., Hartley-Whitaker, J. and Bleeker, P. M., 2002. "The role of phytochelatins in constitutive and adaptive heavy metal tolerances in hyperaccumulator and non-hyperaccumulator metallophytes." *Journal of experimental botany*. 53(379):2381-2392.

Schor-Fumbarov, T., Goldsbrough, P. B., Adam, Z. and Tel-Or, E., 2005. "Characterization and expression of a metallothionein gene in the aquatic fern *Azolla filiculoides* under heavy metal stress." *Planta*, 223:69-76.

Schwitzguébel, J. P., 2001. "Hype or hope: the potential of phytoremediation as an emerging green technology." *Remediation Journal: The Journal of Environmental Cleanup Costs, Technologies & Techniques*, 11(4):63-78.

Sela, M., Fritz, E., Huttermann, A. and Tel-Or, E., 1990. "Studies on cadmium localization in the water fern *Azolla*." *Physiologia Plantarum*, 79(3):547-553.

Sela, M., Garty, J. and Tel-Or, E. 1989. "The accumulation and the effect of heavy metals on the water fern *Azolla filiculoides*." *New Phytologist*, 112(1):7-12.

Sela, M., Tel-Or, E., Fritz, E. and Huttermann, A., 1988. "Localization and toxic effects of cadmium, copper, and uranium in *Azolla*." *Plant Physiology*, 88(1):30-36.

Seregin, I. V. and Kozhevnikova, A. D. 2023. "Phytochelatins: sulfur-containing metal (loid)-chelating ligands in plants." *International Journal of Molecular Sciences*. 24(3):2430.

Shahid, M., Pourrut, B., Dumat, C., Nadeem, M., Aslam, M. and Pinelli, E., 2014. "Heavy-Metal-Induced Reactive Oxygen Species: Phytotoxicity and Physicochemical Changes in Plants." *Reviews of Environmental Contamination and Toxicology* In D. Whitacre, 232:1-44 Springer.

Shanker, A. K., Cervantes, C., Loza-Tavera, H. and Avudainayagam, S., 2005. "Chromium toxicity in plants." *Environment international,* 31(5):739-753.

Sharma, A. and Sachdeva, S., 2015. "Assessing the phytoremediation potential and the metabolic responses of *Azolla microphylla* on lead exposure." *International journal of scientific research*, 4:255-258.

Shedeed, Z. A. and Farahat, E. A., 2023. "Alleviating the toxic effects of Cd and Co on the seed germination and seedling biochemistry of wheat (*Triticum aestivum* L.) using *Azolla pinnata.*" *Environmental Science and Pollution Research*, 1-12.

Singh, J. and Kalamdhad, A. S. 2011. "Effects of heavy metals on soil, plants, human health and aquatic life." *Int J Res Chem Environ.* 1(2):15-21.

Song, Y., Jin, L. and Wang, X., 2017. "Cadmium absorption and transportation pathways in plants." *International journal of phytoremediation* 19(2):133-141.

Sood, A., Uniyal, P. L., Prasanna, R. and Ahluwalia, A. S., 2012. "Phytoremediation potential of aquatic macrophyte, *Azolla.*" *Ambio* 41:122-137.

Spanò, C., Bottega, S., Sorce, C., Bartoli, G. and Ruffini Castiglione, M., 2019. "TiO 2 nanoparticles may alleviate cadmium toxicity in co-treatment experiments on the model hydrophyte *Azolla filiculoides.*" *Environmental Science and Pollution Research*, 26:29872-29882.

Stepniewska, Z., Bennicelli, R. P., Balakhina, T. I., Szajnocha, K., Banach, A. M. and Wolinska, A., 2005. "Potential of *Azolla caroliniana* for the removal of Pb and Cd from wastewaters."

Talebi, M., Tabatabaei, B. E. S. and Akbarzadeh, H., 2019. "Hyperaccumulation of Cu, Zn, Ni, and Cd in *Azolla* species inducing expression of methallothionein and phytochelatin synthase genes." *Chemosphere*, 230:488-497.

Tan, C. Y., Shan, X. Q., Xu, G. Z., Lin, Y. M. and Chen, Z. L., 2011. "Phytoaccumulation of cadmium through *Azolla* from aqueous solution." *Ecological Engineering* 37(11):1942-1946

Tomsett, A. B. and Thurman, D. A. 1988. "Molecular biology of metal tolerances of plants." *Plant, Cell & Environment* 11(5):383-394.

Valderrama, A., Tapia, J., Peñailillo, P. and Carvajal, D. E., 2013. "Water phytoremediation of cadmium and copper using *Azolla filiculoides* L am. in a hydroponic system." *Water and Environment Journal*, 27(3):293-300.

Van Hove C. and Lejeune A. 2002. "The *Azolla*–Anabaena Symbiosis. Biology and Environment," *Proceedings Of The Royal Irish Academy*, 102B(1):23-26.

Van Kempen, M. M., Smolders, A. J., Bögemann, G. M., Lamers, L. L., Visser, E. J. and Roelofs, J. G., 2013. "Responses of the *Azolla filiculoides* Stras.–*Anabaena azollae* Lam. Association to Elevated Sodium Chloride Concentrations: Amino Acids as Indicators for Salt Stress and Tipping Point," *Aquatic Botany*, 106:20-28.

Vaseem, H. and Banerjee, T. K., 2012. "Phytoremediation of the toxic effluent generated during recovery of precious metals from polymetallic sea nodules." *International journal of phytoremediation* 14(5):457-466.

Verbruggen, N., Hermans, C. and Schat, H. 2009. "Molecular mechanisms of metal hyperaccumulation in plants." *New phytologist*. 181(4):759-776.

Veselý, T., Tlustoš, P. and Száková, J., 2011. "The use of water lettuce (*Pistia stratiotes* L.) for rhizofiltration of a highly polluted solution by cadmium and lead." *International journal of phytoremediation*, 13(9):859-872.

Wani, P. A., Khan, M. S. and Zaidi, A., 2008. "Effects of heavy metal toxicity on growth, symbiosis, seed yield and metal uptake in pea grown in metal amended soil." *Bulletin of environmental contamination and toxicology*. 81:152-158.

Williams, J. B. 2002. "Phytoremediation in Wetland Ecosystems: Progress, Problems, and Potential." *Critical Reviews in Plant Sciences* 21(6):607-635.

Xinwei, H., Yongdong, W., Nan, H., Yinhua, S. and Dexin, D., 2018. "Bioremediation of effluent from a uranium mill tailings repository in South China by *Azolla–Anabaena*." *Journal of Radioanalytical and Nuclear Chemistry*, 317:739-746.

Yan, Z. S., Guo, H. Y., Song, T. S., Hu, Y. and Jiang, H. L., 2011. "Tolerance and remedial function of rooted submersed macrophyte *Vallisneria spiralis* to phenanthrene in freshwater sediments." *Ecological Engineering* 37:123-127.

Zazouli, M. A., Mahdavi, Y., Bazrafshan, E. and Balarak, D., 2014. "Phytodegradation potential of bisphenolA from aqueous solution by *Azolla filiculoides*." *Journal of Environmental Health Science and Engineering*, 12:1-5.

Zhang, X., Lin, A. J., Zhao, F. J., Xu, G. Z., Duan, G. L. and Zhu, Y. G., 2008. "Arsenic accumulation by the aquatic fern *Azolla*: comparison of arsenate uptake, speciation and efflux by *A. caroliniana* and *A. filiculoides*." *Environmental Pollution*, 156(3):1149-1155.

Chapter 5

Algal Lipid Metabolism: Recent Advances in Growth and Adaptation

Yigit Sabri Unlu[1]
Hande Morgil[2]
Yusuf Can Gercek[2]
and Taylan Kosesakal[2,*]

[1]Institute of Graduate Studies in Sciences, Istanbul University, Istanbul, Turkey
[2]Botany Division, Department of Biology, Faculty of Science, Istanbul University, Vezneciler, Istanbul, Turkey

Abstract

Microalgae have become the most popular research area in recent years due to their potential in environmental, fuel, food, food supplement and health/cosmetic industries. Microalgaes- unicellular biofactories- have exclusive photosynthetic capacity than terrestrial plants and can adapt to changing conditions. These environmental conditions such as temperature, pH, light, nutrients, gases and organic/inorganic pollutants affect biomass, lipid content and lipid metabolism of microalgae directly or indirectly. Microalgal lipids and their derivatives can be found in many forms for example waxes, terpenoids, fatty acids and phospholipids. In particular, polyunsaturated fatty acids (PUFA) a type of fatty acid are predominantly researched due to their biofuel production potential. Besides of these, docosahexaenoic acid (DHA) and eicosapentaenoic acid (EPA), one of many omega-3 acids and pigments like astaxanthin and B-carotene are also of critical importance due to their effects on health benefits.

---

* Corresponding Author's Email: taylank@istanbul.edu.tr.

In: Advances in Biology. Volume 6
Editor: Charles D. Grant
ISBN: 979-8-89113-712-7

In this section, microalgal products have become sustainable, environmentally friendly resources and that do not compete with agricultural products. However, their efficiencies depend on the environmental conditions and natural or genetically modified strains. Therefore it will be discussed how the lipid amount and biomass of economically important microalgae are affected by environmental conditions. This review critically assesses the physiological, biochemical, and molecular impacts of environmental conditions on microalgal lipids and lipid metabolisms. However, the effects of environmental and nutritional factors as well as stress conditions on lipid metabolism in microalgae will be discussed.

Keywords: microalgae, growth, lipid, lipid metabolism

Introduction

It is very difficult to make a single definition for algae in phycology. In general, algae are aquatic photosynthetic organisms and are represented by 4 kingdoms: Plantae, Protozoa, Bacteria and Chromista (Guiry, 2012). Being systematically represented by 4 kingdoms brings with a large number of species diversity. Algaebase is a database for algae (Guiry et al., 2014) and according to this database, recently 174.086 species and infraspecific names have been registered. Algae are divided into macroalgae and microalgae according to their morphology. For example, red algae and brown algae are considered as a macroalgae; whereas dinoflagellates, cyanobacteria and some green algae are considered as a microalgae (Suganya et al., 2016). Oxygenic microbial photosynthesis evolved 2.4 billion years ago and around 1-2 billion years ago eukaryotic microalgae evolved (Chisti, 2018). Today algae are responsible for 50% of atmospheric oxygen (Chapman, 2013). Chlorophyll is a major pigment used in photosynthesis, there are four types of Chlorophyll: a, b, c and d. Algae contains at least two types of these chlorophylls (Morançais et al., 2018). Withal, their photosynthetic efficiency is 3-4 times higher than terrestrial plants (Anyanwu et al., 2018). Photosynthesis is the primary source of metabolism in algae, high photosynthetic efficiency means high efficiency of primary metabolism products such as carbohydrates, lipids and proteins. Recently, algae have become a popular research area due to their potential. Their products are being used in the environmental, fuel, food, food supplement, health/cosmetic industries. Lipids are the most studied components of algae, Figure 1 shows the number of publications with the keywords "algae" and "lipid" in the Web of Science (WOS) database for the last decade. In this section, we discuss microalgae, microalgal lipids and their lipid metabolisms.

Microalgae act as a biofactory to produce high-value products by using atmospheric carbon dioxide and light (Fernandes and Cordeiro, 2021). They have different shapes and sizes (2-50 μm) and grow in all kinds of aquatic environments (Elisabeth et al., 2021) even in the arctic (Lyon and Mock, 2014) and in hot springs (Ghozzi et al., 2013). Microalgae tolerate harsh conditions such as pollution, pH, light and temperature but their growth rate and primer metabolites or derivatives can be affected. According to their primer metabolite accumulation, microalgae are divided into three classes: protein, carbohydrate and lipid. Table 1 points out the content of primer metabolites produced by important microalgae species (Mutanda et al., 2020). Microorganisms that accumulate more than 20% lipid on a dry cell weight are called oleaginous microorganisms (Cheirsilp et al., 2017). In this case, microalgae that accumulate lipids are called oleaginous microalgae. Lipids can exist in many forms in cells and these microalgal lipids are used in the industry on a broad range.

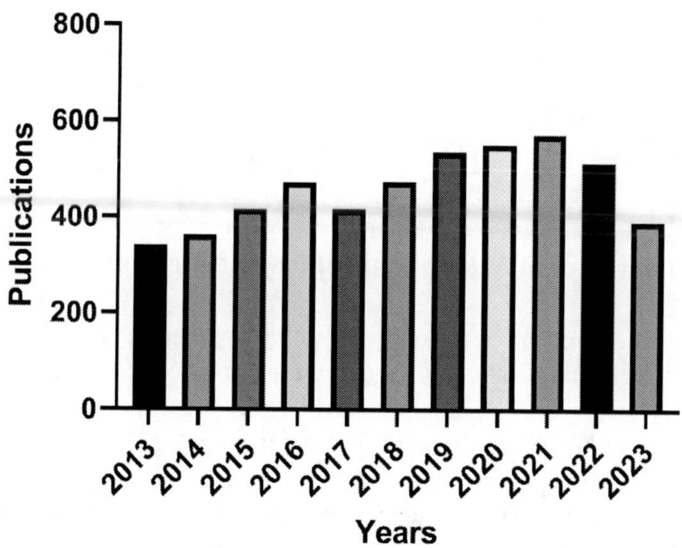

Figure 1. Number of publications of last decade with the keywords "algae" and "lipid" in the Web of Science (WOS) database.

Table 1. Biochemical composition of economically important microalgal strains (% of dry weight) (Mutanda et al., 2020)

Microalgae species	Protein (%)	Carbohydrate (%)	Lipid (%)
Anabaena cylindrical	43-56	25-30	4-7
Botryococcus braunii	-	-	25-75
Chaetoceros calcitrans	58	10	39
Chaetoceros muelleri	44-65	11-19	33
Chlamydomonas rheinhardii	48	17	21
Chlorella protothecoides	10-52	10-15	55
Chlorella pyrenoidosa	57	26	2
Chlorella vulgaris	51-58	12-17	14-22
Chlorella sorokiana	-	-	19-22
Dunaliella bioculata	49	4	8
Dunaliella salina	57	32	6
Euglena gracilis	39-61	14-18	14-20
Heamatococcus pluvialis	-	-	25
Isochyrsis galbana	30-45	7-25	21-38
Phaedactylum tricornutum	18-57	-	-
Porphyridium cruentum	28-39	40-57	9-14
Prymnesium parvum	28-45	25-33	22-38
Scenedesmus obliquus	50-56	35-55	10-17
Scenedesmus dimorphus	8-18	21-52	16-40
Scenedesmus quadricauda	47	-	1-9
Sprigyra sp.	6-20	33-64	11-21
Spirulina maxima	60-71	13-16	6-7
Spirulina platensis	46-63	8-14	4-9
Synechococcus sp.	63	15	11
Schizochytrium sp.	-	-	55-77

Industrial Uses of Microalgal Lipids and Derivates

Microalgae are one of the promising biological resources that show the "5F"s (Harada et al., 2020). The "5F" term represents the food, fiber, feed, fertilizer and fuel (Suzuki, 2017), in addition to this, microalgal compounds are used in cosmetic and health industries. With the growing trend of vegan cosmetics, plant-based cosmetic raw materials are becoming more popular than animal-based cosmetic ingredients (Huynh et al., 2019). Waxes -a class of lipids (Fahy et al., 2005)-, obtained from microalgae are used in lipsticks as a structuring agent (Huynh et al., 2020) and in sunscreen cream to maintain appropriate viscosity (Huynh et al., 2019). pH plays an important role in the physical properties and stability of the lipid membrane in the skin. Fatty acids such as lauric acid, myristic acid, palmitic acid, and stearic acid contribute to skin health by keeping the pH of the skin between 4.0 and 6.5 and these fatty acids can be used as an emulsifier, softener, cleaner and brightener (Yang et al., 2020). Although the amounts vary, microalgae contain these fatty acids (Dolganyuk et al., 2020; Arutselvan et al., 2021; Otero et al., 2017) and microalgal fatty acid usage in cosmetic industries has grown significantly (De Luca et al., 2021). The fatty acids identified in microalgae are summarized in Table 2 (Maltsev and Maltseva, 2021). Furthermore, microalgal extracts are used in skin care, e.g., protein-rich extract from *Arthrospira* or an extract from *Chlorella vulgaris* induces collagen synthesis and promotes tissue regeneration (Spolaore et al., 2006). However, these products are not directly lipid-derived products.

Table 2. Summary of the fatty acids identified in microalgae

Fatty acids	
Polyunsaturated	
5:2 (n.a.)	18:2n-9cis (cis-Δ6,9)
11:2n-3cis (cis-Δ5,8)	18:2cis (cis-Δ8,x)
12:2n-6cis (cis-Δ3,6)	18:2n-6trans (trans-Δ9,12) (Linolelaidic)
12:2n-3cis (cis-Δ6,9)	18:2n-6cis(cis-Δ9,12) (LA)
12:3n-3cis (cis-Δ3,6,9)	18:2n-4cis (cis-Δ9,14)
13:2n-6cis (cis-Δ4,7)	18:2n-4cis (cis-Δ11,14)
14:2n-6cis (cis-Δ5,8)	18:2n-3cis (cis-Δ12,15)
14:3n-3cis (cis-Δ5,8,11)	19:4 (n.a.)
14:5 (n.a.)	20:6(n.a.)
15:4 (n.a.)	20:2n-6cis (cis-Δ11,14) (Eicosadienoic)
15:3n-3cis (cis-Δ6,9,12)	20:3n-6cis (cis-Δ8,11,14) (DGLA)
16:2n-7cis (cis-Δ7,9)	20:3n-3cis (cis-Δ11,14,17) (Eicosatrienoic)
16:2n-7cis (cis-Δ5,9)	20:4n-6cis (cis-Δ5,8,11,14) (ARA)
16:2n-6cis (cis-Δ7,10)	20:4n-5cis (cis-Δ6,9,12,15)
16:2n-4cis (cis-Δ9,12)	20:4n-3cis (cis-Δ8,11,14,17) (ETA)
16:3n-6cis (cis-Δ4,7,10)	20:5n-3cis (cis-Δ5,8,11,14,17) (EPA)
16:3n-4cis (cis-Δ6,9,12)	22:4n-6cis (cis-Δ7,10,13,16) (Adrenic)
16:3n-3cis (cis-Δ7,10,13) (Hexadecatrienoic)	22:5n-6cis (cis-Δ4,7,10,13,16) (Osbond)
16:3n-1cis (cis-Δ9,12,15	22:5n-3cis (cis-Δ7,10,13,16,19) (DPA)
16:4n-3cis (cis-Δ4,7,10,13)	22:6n-3cis (cis-Δ4,7,10,13,16,19) (DHA)
16:4n-1cis (cis-Δ6,9,12,15)	22:6n-6cis (cis-Δx,x,x,x,13,16)
16:4n-4cis (cis-Δx,x,x,12)	28:7n-6cis (cis-Δ4,7,10,13,16,19,22)
17:3n-5cis (cis-Δ6,9,12)	28:8n-3cis (cis-Δ4,7,10,13,16,19,22,25)
17:3n-3cis (cis-Δ8,11,14)	29:3 (n.a.)
17:3n-6cis (cis-Δx,x,11)	
18:4n-3cis (cis-Δ5,9,12,15)	
18:4n-3cis (cis-Δ6,9,12,15) (SDA)	
18:4n-1cis (cis-Δ8,11,14,17)	

Table 2. (Continued)

Fatty acids	
Saturated	**Monounsaturated**
8:0	10:1 (n.a.)
10:0 (Capric)	12:1 (n.a.)
11:0	13:1 (n.a.)
12:0 (Lauric)	14:1n-7cis (cis-Δ7)
13:0	14:1n-6cis (cis-Δ8)
14:0 (Myristic)	14:1n-5cis (cis-Δ9)
14:0 (Myristic)	15:1n-5cis (cis-Δ10)
16:0 (Palmitic)	16:1n-11cis (cis-Δ5)
17:0 (Heptadecanoic)	16:1n-13trans (trans-Δ3)
18:0 (Stearic)	16:1n-9cis (cis-Δ7) (Hypogenic)
19:0	16:1n-8cis (cis-Δ8)
20:0 (Arachidic)	16:1n-7cis (cis-Δ9) (Palmitoleic)
21:0	16:1n-5cis (cis-Δ11)
22:0 (Behenic)	17:1n-8cis (cis-Δ9)
23:0	17:1n-7cis (cis-Δ10)
24:0 (Lignoceric)	18:1n-12cis (cis-Δ6)
27:0	18:1n-9trans (trans-Δ9) (Elaidic)
29:0	18:1n-9cis (cis-Δ9) (Oleic)
	18:1n-7cis (cis-Δ11) (Vaccenic)
	18:1n-5cis (cis-Δ13)
	19:1n-8cis (cis-Δ11)
	20:1n-9cis (cis-Δ11) (Gondoic)
	21:1 (n.a.)
	22:1n-9cis (cis-Δ13) (Erucic)
	23:1 (n.a.)
	24:1n-9cis (cis-Δ15) (Nervonic)

Table 3. Summary of microalgal strains for biofuel production

Microalgae species	References
Chlorella spp.	(Sakarika and Kornaros, 2019); (Das et al., 2018); (Makareviciene et al., 2011); (Rodríguez-Palacio et al., 2022); (kim et al., 2017);
Isochrysis galbana	(Silitonga et al., 2017), (Roopnarain et al., 2014); (Mendoza et al., 2020); (Sánchez-Bayo et al., 2020);
Scenedesmus spp.	(Makareviciene et al., 2011); (Vadivel et al., 2020); (kim et al., 2017);
Neochloris oleoabundans	(Rodríguez-Palacio et al., 2022) (Khan et al., 2023); (Popovich et a. al, 2012); (Levine et al., 2011); (Baldisserotto et al., 2016)
Coelastrella sp.	(Rodríguez-Palacio et al., 2022); (Lee et al., 2021); (Suh et al., 2024); (Minhas et al., 2016)
Verrucodesmus verrucosus	(Rodríguez-Palacio et al., 2022); (Vélez-Landa et al., 2021);
Micractinium inermum	(kim et al., 2017); (Park et al., 2018)
Botryococcus braunii	(Hidalgo et al., 2016); (Tasić et al., 2016); (Prathima and Karthikeyan, 2017);
Nannochloropsis sp.	(Mitra et al., 2015); (Brennan and Regan, 2020); (Rahman et al., 2019); (Li et al., 2011); (Kholssi et al., 2023)
Nitzschia spp.	(Joseph et al., 2017); (Saranya and Ramachandra, 2021); (Kholssi et al., 2023)
Dunaliella spp.	(Patel et al., 2023); (Tang et al., 2011); (Tizvir et al., 2022); (Fawzy and Alharthi, 2021); (Liu et al., 2013)
Desmodesmus spp.	(Okpozu et al., 2019); (Ogbonna et al., 2018);
Chlamydomonas spp.	(Scranton et al., 2015); (Toyoshima and Sato, 2015); (Morowvat et al., 2010); (Cakmak et al., 2012)
Schizochytrium sp.	(Ge et al., 2022); (Chen et al., 2015); (Kuan et al., 2019)

Microalgal fuels have attracted the attention of researchers over two decades. Third-generation biofuels from microalgae are a promising, sustainable source and favorable against first and second-generation biofuels (Chowdhury and Loganathan, 2019). Triacylglycerols (TAG) content of *Chlorella vulgaris* can up to 58% and these contents consist of fatty acids suitable for biodiesel production. Besides *C. vulgaris* is a fast-growing microalgae that is resistant to harsh conditions (Sarayloo et al., 2017 and 2018). There are many potential microalgae species for biodiesel synthesis and these species are summarized in Table 3.

The largest industrial use of microalgae is in the food sector. Several microalgae species (*Chlorella vulgaris*, *Arthrospira platensis*, *Chlamydomonas reinhardtii*, *Dunaliella bardawil*, *Euglena gracilis Auxenochlorella* and *Auxenochlorella protothecoides*) have already been approved by European Food Safety Authority (EFSA) (Molino et al., 2018) and U.S. Food and Drug Administration (FDA) (Torres-Tiji et al., 2020). The use of algae and microalgae as food dates back to BC, historical records show the consumption algae as food dating to 14,000 years ago (Dillehay et al., 2008). Microalgae are rich in protein, carbohydrates and lipids depending on the species and used as a food additive depending on the food type; for example *Spirulina* sp. as additive in breads, biscuits, snacks, desserts, dairy products; *Chlorella* sp. additive in dairy products; *Dunaliella* sp., *Isochrysis galbana* and *Diacronema vlkianum* additive in pastas or just as a cooking oil (Beheshtipour et al., 2013; Kusmayadi et al., 2021; Matos, 2017). The aim here is to improve the quality and the nutritional value of the product along with health benefits. Such as the addition of *Chlorella vulgaris* to wheat flour positively contributes to dough viscosity, the addition of *Haematococcus pluvialis* to cookies is decrease rate of sugar released or the addition of *Dunaliella salina* to pastas make it much more nourishing in terms of polyunsaturated fatty acids (PUFA) (Kusmayadi et al., 2021). Microalgal lipids generally consist of fatty acids with carbon chain lengths ranging from C12-C22, including essential fatty acids such as omega-3 (α-linolenic, eicosapentaenoic, docosahexaenoic acids) and omega-6 (linoleic, γ-linolenic, arachidonic acids) and these fatty acids have a crucial role in human metabolism (Matos, 2017). Besides the consumption of microalgae as food, microalgae are also used as a food supplement to maintain physical and mental health. Microalgae-derived fatty acids in the nutraceutical and pharmaceutical industry are worth $10 billion. The consumption of microalgal supplements can reduce hypertension, thrombosis, cardiovascular diseases and regulate blood cholesterol level. In addition to some fatty acids such as omega-3 and omega-6 fatty acids are necessary for normal growth and functional development in humans and can be used in adjuvant treatment of depression or inflammatory conditions (Adarme-Vega et al., 2012). It is also effective on eye health by regenerating rhodopsin pigments in the retina or regulating the expression of relevant genes (Patel et al., 2020). Microalgal lipids are also rapidly gaining popularity due to pharmaceutical applications such as anticancer agents. One such example is an anticancer agent, a novel compound called "chlorosulfolipid" which is an inhibitor to protein tyrosine kinase (PTK) (Gerwick et al., 1994). PTK is involved in signal transduction leading to cell proliferation and is closely related with cancer (Hubbard and Till, 2000). The intake of omega-3 fatty acids reduces the risk of breast (Zheng et al., 2013) and colorectal cancers (Wu et al., 2012). Microalgae, which have a positive effect on animals as well as being beneficial for humans, are also used in animal feed. According to animal species, we can consider feeds in 3 categories: aqua feed, farm feed and pet food. Aquaculture and livestock farming are the most important sectors together with agriculture for the human population. As the number of people on the planet continues to grow, so does the demand for more production across all food-related industries. Like other animals, fish are unable to produce essential fatty acids and instead accumulate from microalgae (Koller et al., 2014). *Chlorella*, *Tetraselmis*, *Isochrysis*, *Pavlova*, *Phaeodactylum*, *Chaetoceros*, *Nannochloropsis*, *Skeletonema* and *Thalassiosira* are frequently used species in fish feed due to being easily cultured and nontoxic (Spolaore et al., 2006; Levasseur et. al, 2020). *Haematococcus* is a good source for carotenoids and fatty acids. These nutrients play a crucial role in aquaculture especially salmon feed in order to pigmentation and high polyunsaturated fatty acid content (Koller et al., 2014; Spolaore et al., 2006). On the other hand, some studies show that microalgae usage in aquaculture affects fish growth, survivability, immune response, expression of antioxidant genes and fillet quality (Sheikhzadeh et al., 2024). Likewise, the use of microalgal feed on shrimps (de Moraes et al., 2022; Dineshbabu et al., 2019; Huang et al., 2022) and sea urchins (Dineshbabu et al., 2019; Chu et al., 2023; Brundu et al., 2017; Castilla-Gavilán et al., 2018) have positive effects on growth, larvae development, survivability, immune system, meat and water quality, as well as bacterial control of the system. Corn and soybean are the two common feed crops for animal feeding and humans. Land degradation, water and climate crisis are also huge threats for both livestock and agriculture, therefore, microalgae are a substainal aquatic resource (Madeira et al., 2017). Due to their high amount of proteins, lipids and other nutrients microalgae are promising feeds for livestock (Kholif et al.,

2017). By piglets with 5% of their diet containing *Chlorella*, there is no effect on growth although it is to the advantage of the consumer by increasing carotenoid and PUFA levels and improving meat quality (Martins et al., 2021). By goats with a 5-10% *Chlorella*-containing diet affects their milk production and milk composition (Kholif et al., 2017; Kholif et al., 2022). Several studies on lambs, cows and rabbits have demonstrated that microalgae-containing diets increase final body weight and fat levels of animals and PUFA content, decrease cooking loss of meat (Madeira et al., 2017). Eggs are also a valuable protein source, using microalgae in poultry farming has an influence on average daily gain, total carotenoids and PUFA of egg and meat of broilers (Costa et al., 2024; Alfaia et al., 2021; Bruneel et al., 2013). The use of microalgae as feed is versatile, use as pet food have benefits for the health of animals (Dalmonte et al., 2023; Scheibel et al., 2021; Cabrita et al., 2022).

Effects of the Environment on Lipid Metabolism and Growth in Microalgae

For lipid production in microalgae, there are two points to consider: increasing the available biomass or increasing the amount of lipids in the available biomass (Figure 2) (Singh et al., 2016). Therefore, these two factors should be taken into account when investigating the environmental factors affecting the growth, and the lipid metabolism of microalgae. The microalgae culture conditions had been investigated in many previous studies.

Temperature directly affects the lipid content/composition and growth according to a detailed study by Nalley et al., (2018), the optimal growth and lipid content temperature for 26 species belonging to 6 different taxonomic groups ranges between 20-30°C. The parameters vary depending on the industry in which it will be used, for example although some Chlorophyceae members achieved the highest omega-3 fatty acid (α-Linolenic acid) content at 20°C, but the optimal growth rate was achieved at 25°C (Nalley et al., 2018).

Light has a limiting effect on microalgae growth, too intense lighting can enhance the production of reactive oxygen species (ROS) in cells and it may cause photoinhibition, whereas insufficient light slows down the growth rate (Shi et al., 2020). Many studies have sought the optimal light intensity value for microalgae ranging from 5-3500 μmol photons m^{-2} s^{-1} (Maltsev et al., 2021). *C. vulgaris* and *S. obliquus* showed the best growth rate to be at 300 μmol m^{-2} s^{-1} light intensity, however the fatty acid content correlated on *S. obliquus* at increasing light intensity, whereas in *Chlorella* it was higher at 50 μmol m^{-2} s^{-1} and the fatty acid composition was slightly changed (Nzayisenga et al., 2020).

CO_2 and pH both affect the growth of microalgae simultaneously. The level of inorganic carbon in water is in equilibrium as carbon dioxide (CO_2), bicarbonate (HCO_3^-) and carbonate (CO_3^{2-}). Most microalgae can use only CO_2 and HCO_3^- as carbon sources and their distribution levels in water are equal at pH 6 and this equilibrium starts to break down with increasing pH. At pH 9 or above the relative level of CO_3^{-2} increases and CO_3^{-2} is a non-preferred source for microalgal photosynthesis (de Paula Silva et al., 2013). Generally, carbon dioxide (CO_2) above 5% (v/v) is considered toxic to microalgae and can adversely affect growth (Cheah et al., 2015) but some strains have been reported to grow better at 15% (v/v) CO_2 (Kaewkannetra et al., 2012).

Most of the microalgae are autotroph, but microalgae can also be cultivated under mixotrophic conditions. The addition of carbon sources such as glucose, maltose and sodium acetate into the medium induces the growth and lipid content of microalgae (Liu et al., 2021). This is also the case for wastewater containing organic carbon, which can promote the growth of microalgae and have an impact in wastewater treatment (Cho et al., 2011). Gao et al., (2019) showed that microalgae grown in artificial wastewater rich in total organic carbon have higher growth and oil content.

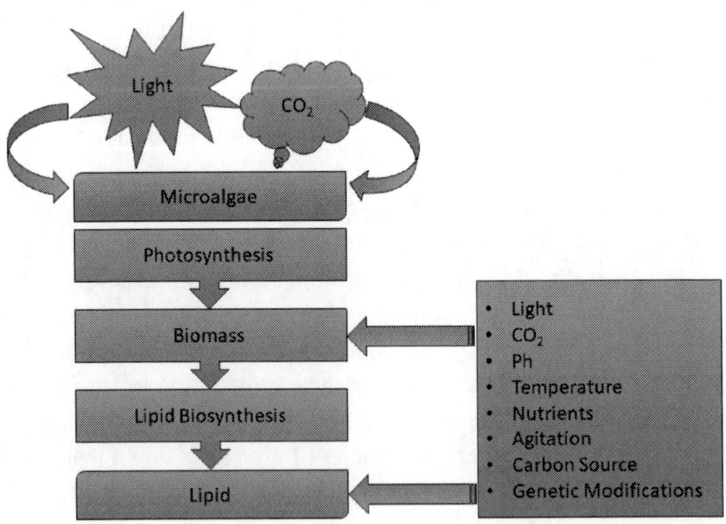

Figure 2. Lipid enhancement strategies for microalgae. Adapted from Singh et al., (2016).

Agitation is another important factor for the development of microalgae due to having equal distribution of gasses, light and nutrients in culture (Bautista-Monroy et al., 2022). Even though, microalgae cell damage can be caused by mechanical agitation of culture systems. This damaging effect is dependent on species, e.g., *Dunaliella* sp. is sensitive to agitation due to lack of cell wall, but *Ochromonas malhamensis* can tolerate agitation up to 350 rpm (Ajala and Alexander, 2022).

Elements have been the most studied factor on the growth and lipid content of microalgae. Elements such as nitrogen, iron, phosphorus, calcium and magnesium are key elements and are involved in physiological pathways. Nitrogen is involved in the structure of proteins, nucleic acids and chlorophyll; therefore it is essential for growth and photosynthesis for microalgae. Phosphorus is also involved in the structure of nucleic acids and various metabolic events. Magnesium is a core element in chlorophyll and iron is responsible for electron transportation during respiration and photosynthesis. The deficiency of magnesium and iron directly affects the growth and the lipid metabolism due to limiting metabolic processes. Finally, calcium is associated with signal transduction and regulates lipid synthesis (da Silva Ferreira and Sant'Anna, 2017). The environmental factors that affect the growth and lipid metabolism of microalgae are a complex set of interrelated relationships. Excess or deficiency of these factors has different effects from species to species. Thus, the optimal conditions for specialized productions such as for the biofuel or food industries are different for each species.

Lipid Metabolism of the Microalgaes

Chlamydomonas reinhardtii serves as a pivotal model organism in microalgal research, primarily due to the comprehensive publication of its genome in 2007 (Merchant et al., 2007). This microalgae has emerged as the subject of extensive investigation, offering unparalleled insights into microalgal lipid studies. However, it is essential to recognize that microalgae encompass a vast taxonomic diversity, and lipid metabolism exhibits variations across different species (Liu and Benning, 2013). Figure 3 summarizes lipid metabolism in microalgae.

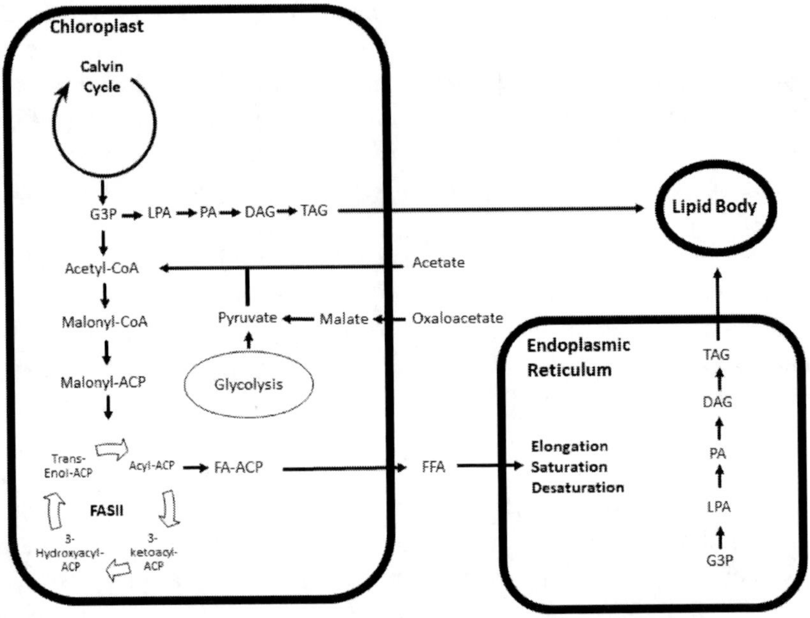

Figure 3. Fatty acid biosynthesis pathway in microalgae. G3P: glycerol 3-phosphate; LPA, lysophosphatidic acid; PA: phosphatidic acid; DAG: diacylglycerol; TAG: triacylglycerol; FFA: free fatty acid; FA-ACP: fatty acid acyl carrier protein. Adapted from Chen & Wang (2021) and Schuhmann & Schenk (2012).

The predominant lipid class in microalgae comprises neutral lipids, predominantly in the form of triacylglycerols (TAG) (Zienkiewicz et al., 2016). Lipid biosynthesis in microalgae predominantly occurs via two pathways: the Kennedy pathway and the fatty acid biosynthesis pathway, operating within the chloroplast and endoplasmic reticulum (ER) (Banerjee et al., 2017). Acetyl-CoA stands as the primary precursor for fatty acid biosynthesis (Shtaida et al., 2015), originating from various cellular compartments such as the mitochondria, chloroplast, and cytoplasm through processes like amino acid catabolism, glycolysis, and β-oxidation (Shi and Tu, 2015; Lin and Oliver, 2008). In the chloroplast, fatty acid synthesis initiates with the conversion of acetyl-CoA to malonyl-CoA catalyzed by acetyl-CoA carboxylase (ACCase). Subsequently, malonyl-CoA undergoes further processing to form 16:0- or 18:0-acyl carrier protein (ACP) via the FAS enzyme complex, serving as essential precursors for TAG synthesis. This process culminates in the storage of lipids as lipid bodies in the cytoplasm, involving the conversion of acyl carrier proteins to diacylglycerol (DAG) and ultimately to TAG (Zienkiewicz et al., 2016; Zulu et al., 2018). Similarly, lipid biosynthesis in the ER, also known as the Kennedy pathway, commences with glycerol-3-phosphate (G3P), which undergoes successive enzymatic reactions to generate TAG via intermediates like lysophosphatidic acid (LPA) and phosphatidic acid (PA) (Chen and Wang, 2021). A schematic representation of lipid biosynthesis, adapted from various studies, elucidates these pathways (Muñoz et al., 2021; Chen and Wang, 2021; Schuhmann et al., 2012; Banerjee et al., 2017).

The fatty acid synthesis pathway, which is ubiquitous among all life forms, is categorized into two types: type 1 (FASI), found in mammals and some lower eukaryotes; and type 2 (FASII), found in bacteria, chloroplasts, and certain microeukaryotes. In essence, the distinction between the two lies in the fact that type 1 is catalyzed by a single enzyme with multiple catalytic sites, whereas in type 2, each step is catalyzed by a separate protein with distinct functions, and microalgae possess a FASII system (Schnorr and Berry, 2022).

Environmental conditions are known to increase lipid content in microalgae, and understanding the underlying molecular mechanisms necessitates omics approaches. Within the fatty acid synthesis pathway in microalgae, several key enzymes play crucial roles. Glycerol-3-phosphate acyltransferase (GPAT), the first enzyme in the Kennedy pathway, is localized in the chloroplast, endoplasmic reticulum (ER), and mitochondria. GPAT catalyzes the conversion of glycerol-3-phosphate (G3P) to lysophosphatidic acid (LPA), thereby providing intermediate molecules necessary for triacylglycerol (TAG) biosynthesis (Duarte-Coello et al., 2019). To date, a few GPAT-like enzymes have been identified in microalgae, and increased expression of GPAT under nitrogen stress has been associated with elevated lipid content (Baral et al., 2012; Misra and Panda, 2013; Xu et al., 2009; Iskandarov et al., 2016; Niu et al., 2016; Duarte-Coello et al., 2019). 2-lysophosphatidic acid acyltransferase (LPAAT), another enzyme in the Kennedy pathway, catalyzes the acylation of LPA (Yamaoka et al., 2016) and interacts with GPAT (Huang et al., 2020). LPAAT has been identified in various microalgae species, and its expression is induced under stress conditions such as light and nitrogen deprivation, correlating with increased lipid content (Sharma et al., 2015; Yamaoka et al., 2016; Gwak et al., 2014). Phosphatidic acid phosphatase (PAP), an enzyme that catalyzes the dephosphorylation of phosphatidic acid (PA) to form diacylglycerol (DAG), remains one of the least studied enzymes. While PAP-homologous genes have been predicted in Chlamydomonas, only PAP2 has a full-length coding sequence, and its expression increases under nitrogen starvation, leading to enhanced lipid content (Merchant et al., 2007; Deng et al., 2013).

Diacylglycerol acyltransferase (DGAT) and phospholipid diacylglycerol acyltransferase (PDAT) are enzymes involved in TAG biosynthesis. DGATs catalyze the final acylation step in TAG synthesis, while PDATs transfer fatty acids from phospholipids to DAG. Both enzymes contribute to TAG production from DAG, with DGATs utilizing phosphatidylcholine (PC) as an acyl donor. While DGAT1 and DGAT2 are well-studied, the role of DGAT3 remains unclear (Guo et al., 2017). Extensive research on lipid metabolism under nitrogen starvation has shown increased lipid content in various microalgae species, correlated with upregulation of DGAT and PDAT expression (Klaitong et al., 2017; Wei et al., 2017; Guo et al., 2017; Úbeda-Mínguez et al., 2017; La Russa et al., 2012; Cui et al., 2018; Wagner et al., 2010).

Acetyl-CoA, the main precursor of fatty acid biosynthesis, is synthesized by acetyl-CoA carboxylase (ACCase), which catalyzes the carboxylation of acetyl-CoA to malonyl-CoA. ACCase plays a crucial role in channeling carbon flux into fatty acid synthesis and has been implicated in lipid metabolism in various microalgae species under different environmental conditions (Li et al., 2018; Ma et al., 2019; Zhang et al., 2021; Singh et al., 2017). Malonyl-CoA ACP transacylase (MAT) and 3-ketoacyl-ACP synthase (KAS) are enzymes involved in the condensation of malonyl-CoA and subsequent formation of 3-ketoacyl-ACP, contributing to lipid biosynthesis under salinity or nitrogen stress in microalgae (Atikij et al., 2019; Guarnieri et al., 2011). Malic enzyme (ME) converts malate to pyruvate, a precursor of acetyl-CoA, and has been implicated in lipid biosynthesis in various microalgae species under different stress conditions (Kim et al., 2019; Shang et al., 2012; Xue et al., 2015; Fan et al., 2014). Additionally, acetyl-CoA synthetase (ACS) catalyzes the conversion of acetate to acetyl-CoA, a key precursor of lipid biosynthesis, and its overexpression under stress conditions promotes TAG biosynthesis in microalgae (Rengel et al., 2018; Liang et al., 2016; Avidan and Pick, 2015; Wu et al., 2019).

Omics Studies of Microalgae

Microalgae, due to their ecological significance, require detailed information about their fundamental biological functions. In recent years, the growing use of "omics" technologies has begun to meet the need for in-depth knowledge about microalgae. The use of omics technologies, collectively referred to as "algomics," allows for the understanding of the cellular blueprint of algal systems, assembling DNA sequences, and analyzing gene structures and expressions (Lauritano et al., 2019). The concept of algomics has evolved to encompass genomic, transcriptomic, proteomic, metabolomic, glycomic, and lipidomic aspects with advancing technologies. The integrated use of algomic approaches can assist in targeting microalgae that can be developed through genetic engineering applications. In addition, omics approaches and the large datasets, which can be analyzed using computer tools and software, can be crucial tools in identifying specific metabolic pathways. Particularly, they have begun to be used in applications such as the production of lipid-based microalgal products and bioactive component, biofuel production, CO_2 sequestration, and wastewater treatment. The importance of microalgae in the quest for biofuels by humanity cannot be underestimated, especially with the growing world population. The use of evolving omics technologies allows for a better understanding of metabolic processes and genetic information, enabling the stimulation of lipid synthesis in microalgae.

To date, more than 60 microalgal species, such as *Chlamydomonas reinhardtii*, *Ostreococcus lucimarinus*, *Dunaliella salina*, *Chlorella vulgaris*, *Volvox carteri* have been fully sequenced and are available in databases (Salama et al., 2019). The sequencing information in these databases is illuminating for understanding the functioning of microalgal cellular pathways, especially in studies related to lipid metabolism. Genomic sequence information provides insights into the metabolic pathways of related organisms and allows for their reorganization. Phylogenetic analyses of gene families in between microalgae species can reveal species-specific metabolic features. Genomic comparisons providing data for genetic engineering studies can be utilized to generate gene information related to the phenotype, aiming to enhance the efficiency of specific molecules. Lipid-based microalgal products can be induced by abiotic stresses such as pH, light, salt, UV radiation, temperature, etc. Characterization of bio-products under these changing environmental conditions is facilitated through genomic studies (Arora et al., 2018).

Transcriptomic studies provide detailed information about the pool of expressed genes, offering deeper insights into cellular functions. Specifically, determining the organism's responses to changing environmental conditions at the genetic level contributes to lipid-based microalgal products research. Guarnieri et al., (2011) demonstrated that nitrogen depletion led to up to a 50% increase in FAME content in *C. vulgaris*. Another study with *Neodesmus sp.*

identified triose phosphate isomerase as a key enzyme regulating the mechanism between fatty acid and starch biosynthesis. Similarly, overexpression of the DGAT-2 gene in *P. tricornutum* and the NoD12 gene in *N. oceanica* increased lipid production (Niu et al., 2013; Kaye et al., 2015). As is known, transcription factors are proteins that, by binding to a nearby DNA location, "activate" or "deactivate" specific genes (Latchman, 1997). Identifying transcription factors in algal genomes has become increasingly important in understanding the regulatory mechanisms of lipid-based microalgal products. Determining transcription factors contributing to lipid biosynthesis and modifying them through genetic engineering can positively influence lipid accumulation in organisms. A novel transcription factor, MYB1, has been identified in *Chlamydomonas*, and under standard growth conditions, overexpression (MYB1-OE) transformants exhibited 1.9 to 3.2 times more triacylglycerol accumulation compared to WT (Zhao et al., 2023). In addition to transcription factors, more research is needed for RNAi-mediated (double-stranded RNAs, small interfering (si)RNAs, and artificial miRNAs) omics studies in algae (Salama et al., 2019).

The identification of protein families involved in lipid increase requires the utilization of proteomic tools. The concept of proteomics, providing insights into gene expression and regulation, should be considered in parallel with genomic and transcriptomic information. Proteomic tools, especially those used to determine the localization of post-translational protein modifications in the cell, can be crucial for microalgal research. Detailed proteomic analyses were conducted to elucidate the mechanisms regulating carbon partitioning in *N. oculata* under nitrogen deprivation (Tran et al., 2016). The organism selectively up- or down-regulated proteins to regulate metabolic flux under low nitrate conditions, resulting in an approximately 60% increase in the amount of fatty acid methyl esters.

Metabolomics is a branch of biotechnology that examines the chemical fingerprints specific to certain biochemical processes, particularly the profiles of all small molecules (metabolites) produced by these processes. Lipids, polysaccharides, polyunsaturated fatty acids, polyphenols, peptides, vitamins, and pigments are components of the microalgal product range. In recent studies, the positive contribution of changing environmental conditions to the rate and quantity of metabolite production has been noticeable (Valledor et al., 2014). In particular, the limitation of nitrogen in the nutrient medium, resulting in an increase in lipid and carbohydrate stores, has been supported by many studies (Valledor et al., 2014; Lin et al., 2011). When microalgal cells are cultivated under nutrient deficiency conditions (P, N, Fe, or S deficiency), they exhibit diverse metabolic profiles, especially resulting in increased lipid content. It is widely known that abiotic stresses have a negative impact on microalgal physiology (Gu et al., 2012). In a study conducted by Saadaoui et al., (2024), it was determined that under high salinity and N-deprivation stress, triglycerides containing α-linoleic (18:3), linoleic (18:2), and palmitic (16:0) fatty acids were prevalent in algae. Overall, when compared with all applied stress factors, high salinity was found to enhance the overall metabolic profile in *T. subcordiformis*, containing positive approaches in lipid metabolism profiling. For a better understanding of lipid metabolism profiles and the identification of positive approaches, more species-specific metabolomic studies involving combined analysis of different stresses need to be included in the literature.

The knowledge and datasets obtained in the field of "algomics" are rapidly increasing today. More research is needed for genetic engineering-based studies by bringing together all the data obtained from omics approaches. Considering the big datasets in databases, using bioinformatics tools can eliminate issues such as the increased cost and excessive time requirements created by traditional research techniques (Cunha et al., 2023). The big data information from omics approaches has recently begun to be interpreted through artificial intelligence with a systems biology perspective (Cunha et al., 2023). Mathematical models simulating cell metabolism under different environmental and genetic conditions will soon form the basis of scientific studies. These models, which combine multiple approaches, are in the early stages for algal studies.

Conclusion

In algae research, leveraging omics techniques is crucial. Omics provides a holistic approach to understanding organisms, offering insights into intricate cellular structures and biological characteristics. Integrating data from various omics methodologies is imperative for a comprehensive analysis, leading to a deeper understanding of the entire system and concurrent biological changes. Moreover, it's possible to develop systems for enhancing the production of specific materials. High-efficiency engineering of algal strains enables improvements in biomass and lipid efficiency across different species, even without prior knowledge of gene functions. Recent advancements in CRISPR technology expedite the creation of precise mutations, strengthening the resilience of high-efficiency mutant libraries and advancing our understanding of algal metabolism. Omics-informed genome-scale models play a pivotal role in systematic species engineering.

Microalgae-based studies are also being conducted in the field of wastewater bioremediation. Prioritizing treatment technologies and nutrient recovery in biomass form is essential for the safe discharge of wastewater and achieving circular economy objectives. Algae are favored for their rapid growth rates and sustainable production of fine chemicals and biofuels. Furthermore, bioactive compounds such as polyunsaturated fatty acids, peptides and pigments can be extracted, purified, and formulated into commercial products. To effectively utilize high-value lipids, it's imperative to enhance our understanding of novel, straightforward, and economical approaches to boost their production. Moreover, employing high-throughput techniques capable of identifying and quantifying various lipid components across a broad spectrum is essential.

Acknowledgments

This study was funded by Scientific Research Projects Coordination Unit of Istanbul University Project number: FBA-2024-40473. The authors wish to thank Cemre Şenesen, for grammatical revision of this chapter.

Disclaimer

None.

References

Adarme-Vega, T. C., Lim, D. K., Timmins, M., Vernen, F., Li, Y., & Schenk, P. M. (2012). Microalgal biofactories: a promising approach towards sustainable omega-3 fatty acid production. *Microbial cell factories*, *11*(1), 1-10.

Ajala, S. O., & Alexander, M. L. (2022). Evaluating the effects of agitation by shaking, stirring and air sparging on growth and accumulation of biochemical compounds in microalgae cells. *Biofuels*, *13*(3), 371-381.

Alfaia, C. M., Pestana, J. M., Rodrigues, M., Coelho, D., Aires, M. J., Ribeiro, D. M., ... & Prates, J. A. M. (2021). Influence of dietary Chlorella vulgaris and carbohydrate-active enzymes on growth performance, meat quality and lipid composition of broiler chickens. *Poultry science*, *100*(2), 926-937.

Anyanwu, R., Rodriguez, C., Durrant, A., & Olabi, A. G. (2018). Micro-macroalgae properties and applications. In *Reference module in materials science and materials engineering*. Elsevier BV.

Arora, N., Pienkos, P. T., Pruthi, V., Poluri, K. M., & Guarnieri, M. T. (2018). Leveraging algal omics to reveal potential targets for augmenting TAG accumulation. *Biotechnology advances*, *36*(4), 1274-1292.

Arutselvan, C., Narchonai, G., Pugazhendhi, A., LewisOscar, F., & Thajuddin, N. (2021). Evaluation of microalgal strains and microalgal consortium for higher lipid productivity and rich fatty acid profile towards sustainable biodiesel production. *Bioresource Technology*, *339*, 125524.

Atikij, T., Syaputri, Y., Iwahashi, H., Praneenararat, T., Sirisattha, S., Kageyama, H., & Waditee-Sirisattha, R. (2019). Enhanced lipid production and molecular dynamics under salinity stress in green microalga Chlamydomonas reinhardtii (137C). *Marine drugs*, *17*(8), 484.

Avidan, O., & Pick, U. (2015). Acetyl-CoA synthetase is activated as part of the PDH-bypass in the oleaginous green alga Chlorella desiccata. *Journal of Experimental Botany*, *66*(22), 7287-7298.

Baldisserotto, C., Popovich, C., Giovanardi, M., Sabia, A., Ferroni, L., Constenla, D., ... & Pancaldi, S. (2016). Photosynthetic aspects and lipid profiles in the mixotrophic alga Neochloris oleoabundans as useful parameters for biodiesel production. *Algal Research*, *16*, 255-265.

Banerjee, A., Maiti, S. K., Guria, C., & Banerjee, C. (2017). Metabolic pathways for lipid synthesis under nitrogen stress in Chlamydomonas and Nannochloropsis. *Biotechnology letters*, *39*, 1-11.

Baral, M., Misra, N., Panda, P. K., & Thirunavoukkarasu, M. (2012). In silico structural determination of GPAT enzyme from Ostreococcus lucimarinus for biotechnological application of microalgal biofuel production. *Biotechnology & Biotechnological Equipment*, *26*(1), 2794-2800.

Bautista-Monroy, S. S., Chávez-Urbiola, E. A., Ortega-Palacios, R., González-Sánchez, A., Gómez-Aldapa, C. A., Rodríguez-Nava, O., ... & Cadena-Ramírez, A. (2022). Insights of Raceway Bioreactor Scale-Up: Effect of Agitation on Microalgae Culture and Reduction of the Liquid Medium Speed. *Applied Sciences*, *12*(3), 1513.

Beheshtipour, H., Mortazavian, A. M., Mohammadi, R., Sohrabvandi, S., & Khosravi-Darani, K. (2013). Supplementation of Spirulina platensis and Chlorella vulgaris algae into probiotic fermented milks. *Comprehensive Reviews in Food Science and Food Safety*, *12*(2), 144-154.

Brennan, B., & Regan, F. (2020). In-situ lipid and fatty acid extraction methods to recover viable products from Nannochloropsis sp. *Science of the Total Environment*, *748*, 142464.

Brundu, G., Vallainc, D., Baroli, M., Figus, A. M., Pinna, A., & Carboni, S. (2017). Effects of on-demand feeding on sea urchin larvae (Paracentrotus lividus; Lamarck, 1816), development, survival and microalgae utilization. *Aquaculture Research*, *48*(4), 1550-1560.

Bruneel, C., Lemahieu, C., Fraeye, I., Ryckebosch, E., Muylaert, K., Buyse, J., & Foubert, I. (2013). Impact of microalgal feed supplementation on omega-3 fatty acid enrichment of hen eggs. *Journal of functional foods*, *5*(2), 897-904.

Cabrita, A. R., Guilherme-Fernandes, J., Valente, I. M., Almeida, A., Lima, S. A., Fonseca, A. J., & Maia, M. R. (2022). Nutritional composition and untargeted metabolomics reveal the potential of Tetradesmus obliquus, Chlorella vulgaris and Nannochloropsis oceanica as valuable nutrient sources for dogs. *Animals*, *12*(19), 2643.

Cakmak, T., Angun, P., Demiray, Y. E., Ozkan, A. D., Elibol, Z., & Tekinay, T. (2012). Differential effects of nitrogen and sulfur deprivation on growth and biodiesel feedstock production of Chlamydomonas reinhardtii. *Biotechnology and bioengineering*, *109*(8), 1947-1957.

Castilla-Gavilán, M., Buzin, F., Cognie, B., Dumay, J., Turpin, V., & Decottignies, P. (2018). Optimising microalgae diets in sea urchin Paracentrotus lividus larviculture to promote aquaculture diversification. *Aquaculture*, *490*, 251-259.

Chapman, R. L. (2013). Algae: the world's most important "plants"—an introduction. *Mitigation and Adaptation Strategies for Global Change*, *18*, 5-12.

Cheah, W. Y., Show, P. L., Chang, J. S., Ling, T. C., & Juan, J. C. (2015). Biosequestration of atmospheric CO2 and flue gas-containing CO2 by microalgae. *Bioresource technology*, *184*, 190-201.

Cheirsilp, B., Thawechai, T., & Prasertsan, P. (2017). Immobilized oleaginous microalgae for production of lipid and phytoremediation of secondary effluent from palm oil mill in fluidized bed photobioreactor. *Bioresource Technology*, *241*, 787-794.

Chen, H., & Wang, Q. (2021). Regulatory mechanisms of lipid biosynthesis in microalgae. *Biological Reviews*, *96*(5), 2373-2391.

Chen, W., Ma, L., Zhou, P. P., Zhu, Y. M., Wang, X. P., Luo, X. A., ... & Yu, L. J. (2015). A novel feedstock for biodiesel production: the application of palmitic acid from Schizochytrium. *Energy*, *86*, 128-138.

Chisti, Y. (2018). Society and microalgae: Understanding the past and present. In *Microalgae in health and disease prevention* (pp. 11-21). Academic Press.

Cho, S. J., Lee, D. H., Luong, T. T., Park, S. R., Oh, Y. K., & Lee, T. H. (2011). Effects of carbon and nitrogen sources on fatty acid contents and composition in the green microalga, Chlorella sp. 227. *Journal of microbiology and biotechnology*, *21*(10), 1073-1080.

Chowdhury, H., & Loganathan, B. (2019). Third-generation biofuels from microalgae: a review. *Current opinion in green and sustainable chemistry*, *20*, 39-44.

Chu, Y., Ding, D. S., Sun, W. T., Satuito, C. G., & Pan, C. H. (2023). Effects of Marine Microalgae on the Developmental Growth of the Sea Urchin Larviculture Anthocidaris crassipina. *Fishes*, *8*(6), 278.

Costa, M. M., Spínola, M. P., & Prates, J. A. (2024). Microalgae as an Alternative Mineral Source in Poultry Nutrition. *Veterinary Sciences*, *11*(1), 44.

Cui, Y., Zhao, J., Wang, Y., Qin, S., & Lu, Y. (2018). Characterization and engineering of a dual-function diacylglycerol acyltransferase in the oleaginous marine diatom Phaeodactylum tricornutum. *Biotechnology for Biofuels*, *11*, 1-13.

Cunha, E., Sousa, V., Geada, P., Teixeira, J. A., Vicente, A. A., & Dias, O. (2023). Systems biology's role in leveraging microalgal biomass potential: Current status and future perspectives. *Algal Research, 69*, 102963.

da Silva Ferreira, V., & Sant'Anna, C. (2017). The effect of physicochemical conditions and nutrient sources on maximizing the growth and lipid productivity of green microalgae. *Phycological Research, 65*(1), 3-13.

Dalmonte, T., Vecchiato, C. G., Biagi, G., Fabbri, M., Andreani, G., & Isani, G. (2023). Iron Bioaccessibility and Speciation in Microalgae Used as a Dog Nutrition Supplement. *Veterinary Sciences, 10*(2), 138.

Das, S. K., Sathish, A., & Stanley, J. (2018). Production of biofuel and bioplastic from Chlorella pyrenoidosa. *Materials today: proceedings, 5*(8), 16774-16781.

De Luca, M., Pappalardo, I., Limongi, A. R., Viviano, E., Radice, R. P., Todisco, S., ... & Vassallo, A. (2021). Lipids from microalgae for cosmetic applications. *Cosmetics, 8*(2), 52.

de Moraes, L. B. S., Santos, R. F. B., Gonçalves Junior, G. F., Mota, G. C. P., Dantas, D. M. D. M., de Souza Bezerra, R., & Olivera Gálvez, A. (2022). Microalgae for feeding of penaeid shrimp larvae: an overview. *Aquaculture International, 30*(3), 1295-1313.

de Paula Silva, P. H., Paul, N. A., de Nys, R., & Mata, L. (2013). Enhanced production of green tide algal biomass through additional carbon supply. *PloS one, 8*(12), e81164.

Deng, X. D., Cai, J. J., & Fei, X. W. (2013). Involvement of phosphatidate phosphatase in the biosynthesis of triacylglycerols in Chlamydomonas reinhardtii. *Journal of Zhejiang University SCIENCE B, 14*, 1121-1131.

Dillehay, T. D., Ramírez, C., Pino, M., Collins, M. B., Rossen, J., & Pino-Navarro, J. D. (2008). Monte Verde: seaweed, food, medicine, and the peopling of South America. *science, 320*(5877), 784-786.

Dineshbabu, G., Goswami, G., Kumar, R., Sinha, A., & Das, D. (2019). Microalgae–nutritious, sustainable aqua-and animal feed source. *Journal of Functional Foods, 62*, 103545.

Dolganyuk, V., Andreeva, A., Budenkova, E., Sukhikh, S., Babich, O., Ivanova, S., ... & Ulrikh, E. (2020). Study of morphological features and determination of the fatty acid composition of the microalgae lipid complex. *Biomolecules, 10*(11), 1571.

Duarte-Coello, M. E., Herrera-Valencia, V. A., Echevarría-Machado, I., Casais-Molina, M. L., & Peraza-Echeverria, S. (2019). Molecular cloning and functional characterization of two glycerol-3-phosphate acyltransferases from the green microalga Chlamydomonas reinhardtii. *Phycological Research, 67*(2), 102-111.

Elisabeth, B., Rayen, F., & Behnam, T. (2021). Microalgae culture quality indicators: a review. *Critical reviews in biotechnology, 41*(4), 457-473.

Fahy, E., Subramaniam, S., Brown, H. A., Glass, C. K., Merrill, A. H., Murphy, R. C., ... & Dennis, E. A. (2005). A comprehensive classification system for lipids1. *Journal of lipid research, 46*(5), 839-861.

Fan, J., Cui, Y., Wan, M., Wang, W., & Li, Y. (2014). Lipid accumulation and biosynthesis genes response of the oleaginous Chlorella pyrenoidosa under three nutrition stressors. *Biotechnology for biofuels, 7*, 1-14.

Fawzy, M. A., & Alharthi, S. (2021). Use of response surface methodology in optimization of biomass, lipid productivity and fatty acid profiles of marine microalga Dunaliella parva for biodiesel production. *Environmental Technology & Innovation*, *22*, 101485.

Fernandes, T., & Cordeiro, N. (2021). Microalgae as sustainable biofactories to produce high-value lipids: Biodiversity, exploitation, and biotechnological applications. *Marine Drugs*, *19*(10), 573.

Gao, F., Yang, H. L., Li, C., Peng, Y. Y., Lu, M. M., Jin, W. H., ... & Guo, Y. M. (2019). Effect of organic carbon to nitrogen ratio in wastewater on growth, nutrient uptake and lipid accumulation of a mixotrophic microalgae Chlorella sp. *Bioresource technology*, *282*, 118-124.

Ge, S., Pugazhendhi, A., Sekar, M., Xia, C., Elfasakhany, A., Brindhadevi, K., & Whangchai, K. (2022). PM emissions-assessment of combustion energy transfer with Schizochytrium sp. algal biodiesel and blends in IC engine. *Science of The Total Environment*, *802*, 149750.

Gerwick, W. H., Roberts, M. A., Proteau, P. J., & Chen, J. L. (1994). Screening cultured marine microalgae for anticancer-type activity. *Journal of Applied Phycology*, *6*, 143-149.

Ghozzi, K., Zemzem, M., Dhiab, R. B., Challouf, R., Yahia, A., Omrane, H., & Ouada, H. B. (2013). Screening of thermophilic microalgae and cyanobacteria from Tunisian geothermal sources. *Journal of arid environments*, *97*, 14-17.

Gu, N., Lin, Q., Li, G., Qin, G., Lin, J., & Huang, L. (2012). Effect of salinity change on biomass and biochemical composition of Nannochloropsis oculata. *Journal of the world aquaculture society*, *43*(1), 97-106.

Guarnieri, M. T., Nag, A., Smolinski, S. L., Darzins, A., Seibert, M., & Pienkos, P. T. (2011). Examination of triacylglycerol biosynthetic pathways via de novo transcriptomic and proteomic analyses in an unsequenced microalga. *PloS one*, *6*(10), e25851.

Guiry, M. D. (2012). How many species of algae are there?. *Journal of phycology*, *48*(5), 1057-1063.

Guiry, M. D., Guiry, G. M., Morrison, L., Rindi, F., Miranda, S. V., Mathieson, A. C., ... & Garbary, D. J. (2014). AlgaeBase: an on-line resource for algae. *Cryptogamie, Algologie*, *35*(2), 105-115.

Guo, X., Fan, C., Chen, Y., Wang, J., Yin, W., Wang, R. R., & Hu, Z. (2017). Identification and characterization of an efficient acyl-CoA: diacylglycerol acyltransferase 1 (DGAT1) gene from the microalga Chlorella ellipsoidea. *BMC Plant Biology*, *17*(1), 1-16.

Gwak, Y., Hwang, Y. S., Wang, B., Kim, M., Jeong, J., Lee, C. G., ... & Jin, E. (2014). Comparative analyses of lipidomes and transcriptomes reveal a concerted action of multiple defensive systems against photooxidative stress in Haematococcus pluvialis. *Journal of experimental botany*, *65*(15), 4317-4334.

Harada, R., Nomura, T., Yamada, K., Mochida, K., & Suzuki, K. (2020). Genetic engineering strategies for Euglena gracilis and its industrial contribution to sustainable development goals: A review. *frontiers in Bioengineering and Biotechnology*, *8*, 790.

Hidalgo, P., Ciudad, G., & Navia, R. (2016). Evaluation of different solvent mixtures in esterifiable lipids extraction from microalgae Botryococcus braunii for biodiesel production. *Bioresource Technology*, *201*, 360-364.

Huang, C., Luo, Y., Zeng, G., Zhang, P., Peng, R., Jiang, X., & Jiang, M. (2022). Effect of adding microalgae to whiteleg shrimp culture on water quality, shrimp development and yield. *Aquaculture Reports*, *22*, 100916.

Huang, L., Yu, L., Li, Z., Li, Y., Yoon, K. S., Hu, Q., ... & Han, D. (2020). Microalgal plastidial lysophosphatidic acid acyltransferase interacts with upstream glycerol-3-phosphate acyltransferase and defines its substrate selectivity via the two transmembrane domains. *Algal Research*, *45*, 101758.

Hubbard, S. R., & Till, J. H. (2000). Protein tyrosine kinase structure and function. *Annual review of biochemistry*, *69*(1), 373-398.

Huynh, A., Abou-Dahech, M. S., Reddy, C. M., O'Neil, G. W., Chandler, M., & Baki, G. (2019). Alkenones, a renewably sourced, biobased wax as an SPF booster for organic sunscreens. *Cosmetics*, *6*(1), 11.

Huynh, A., Maktabi, B., Reddy, C. M., O'Neil, G. W., Chandler, M., & Baki, G. (2020). Evaluation of alkenones, a renewably sourced, plant-derived wax as a structuring agent for lipsticks. *International Journal of Cosmetic Science*, *42*(2), 146-155.

Iskandarov, U., Sitnik, S., Shtaida, N., Didi-Cohen, S., Leu, S., Khozin-Goldberg, I., ... & Boussiba, S. (2016). Cloning and characterization of a GPAT-like gene from the microalga Lobosphaera incisa (Trebouxiophyceae): overexpression in Chlamydomonas reinhardtii enhances TAG production. *Journal of applied phycology*, *28*, 907-919.

Joseph, M. M., Renjith, K. R., John, G., Nair, S. M., & Chandramohanakumar, N. (2017). Biodiesel prospective of five diatom strains using growth parameters and fatty acid profiles. *Biofuels*, *8*(1), 81-89.

Kaewkannetra, P., Enmak, P., & Chiu, T. (2012). The effect of CO 2 and salinity on the cultivation of Scenedesmus obliquus for biodiesel production. *Biotechnology and bioprocess engineering*, *17*, 591-597.

Khan, T. A., Khan, T. A., Yadav, A. K., Khan, M. E., Kumar, M., & Srivastava, P. K. (2023). Experimental studies on utilization of Neochloris oleoabundans microalgae biodiesel as an alternative fuel for diesel engine. *International Journal of Ambient Energy*, *44*(1), 115-123.

Kholif, A. E., Gouda, G. A., Abu Elella, A. A., & Patra, A. K. (2022). Replacing the concentrate feed mixture with Moringa oleifera leaves silage and chlorella vulgaris microalgae mixture in diets of damascus goats: lactation performance, nutrient utilization, and ruminal fermentation. *Animals*, *12*(12), 1589.

Kholif, A. E., Morsy, T. A., Matloup, O. H., Anele, U. Y., Mohamed, A. G., & El-Sayed, A. B. (2017). Dietary Chlorella vulgaris microalgae improves feed utilization, milk production and concentrations of conjugated linoleic acids in the milk of Damascus goats. *The Journal of Agricultural Science*, *155*(3), 508-518.

Kholssi, R., Riouchi, O., Douhri, H., & Debdoubi, A. (2023). Microalgae as a sustainable energy source: Growth and lipids production of Nitzschia sp., Nannochloropsis sp., and Tetraselmis sp. from Mediterranean seawater. *Biocatalysis and Agricultural Biotechnology*, *50*, 102720.

Kim, G. Y., Yun, Y. M., Shin, H. S., & Han, J. I. (2017). Cultivation of four microalgae species in the effluent of anaerobic digester for biodiesel production. *Bioresource technology*, *224*, 738-742.

Kim, J., Kwak, H. S., Sim, S. J., & Jin, E. (2019). Overexpression of malic enzyme isoform 2 in Chlamydomonas reinhardtii PTS42 increases lipid production. *Bioresource Technology Reports*, *7*, 100239.

Klaitong, P., Fa-Aroonsawat, S., & Chungjatupornchai, W. (2017). Accelerated triacylglycerol production and altered fatty acid composition in oleaginous microalga Neochloris oleoabundans by overexpression of diacylglycerol acyltransferase 2. *Microbial Cell Factories*, *16*, 1-10.

Koller, M., Muhr, A., & Braunegg, G. (2014). Microalgae as versatile cellular factories for valued products. *Algal research*, *6*, 52-63.

Kuan, D., Dai, L., Liu, D., Liu, H., & Du, W. (2019). Efficient Biodiesel Conversion from Microalgae Oil of Schizochytrium sp. *Catalysts*, *9*(4), 341.

Kusmayadi, A., Leong, Y. K., Yen, H. W., Huang, C. Y., & Chang, J. S. (2021). Microalgae as sustainable food and feed sources for animals and humans–biotechnological and environmental aspects. *Chemosphere*, *271*, 129800.

La Russa, M., Bogen, C., Uhmeyer, A., Doebbe, A., Filippone, E., Kruse, O., & Mussgnug, J. H. (2012). Functional analysis of three type-2 DGAT homologue genes for triacylglycerol production in the green microalga Chlamydomonas reinhardtii. *Journal of biotechnology*, *162*(1), 13-20.

Latchman, D. S. (1997). Transcription factors: an overview. *The international journal of biochemistry & cell biology*, *29*(12), 1305-1312.

Lauritano, C., Ferrante, M. I., & Rogato, A. (2019). Marine natural products from microalgae: An-omics overview. *Marine drugs*, *17*(5), 269.

Lee, S. A., Ko, S. R., Lee, N., Lee, J. W., Van Le, V., Oh, H. M., & Ahn, C. Y. (2021). Two-step microalgal (Coelastrella sp.) treatment of raw piggery wastewater resulting in higher lipid and triacylglycerol levels for possible production of higher-quality biodiesel. *Bioresource Technology*, *332*, 125081.

Levasseur, W., Perré, P., & Pozzobon, V. (2020). A review of high value-added molecules production by microalgae in light of the classification. *Biotechnology advances*, *41*, 107545.

Levine, R. B., Costanza-Robinson, M. S., & Spatafora, G. A. (2011). Neochloris oleoabundans grown on anaerobically digested dairy manure for concomitant nutrient removal and biodiesel feedstock production. *Biomass and Bioenergy*, *35*(1), 40-49.

Li, D. W., Xie, W. H., Hao, T. B., Cai, J. X., Zhou, T. B., Balamurugan, S., ... & Li, H. Y. (2018). Constitutive and chloroplast targeted expression of acetyl-CoA carboxylase in oleaginous microalgae elevates fatty acid biosynthesis. *Marine Biotechnology*, *20*, 566-572.

Li, Y., Lian, S., Tong, D., Song, R., Yang, W., Fan, Y., ... & Hu, C. (2011). One-step production of biodiesel from Nannochloropsis sp. on solid base Mg–Zr catalyst. *Applied Energy*, *88*(10), 3313-3317.

Liang, M. H., Qv, X. Y., Jin, H. H., & Jiang, J. G. (2016). Characterization and expression of AMP-forming Acetyl-CoA Synthetase from Dunaliella tertiolecta and its response to nitrogen starvation stress. *Scientific reports*, *6*(1), 23445.

Lin, M., & Oliver, D. J. (2008). The role of acetyl-coenzyme a synthetase in Arabidopsis. *Plant physiology*, *147*(4), 1822-1829.

Lin, Q., & Lin, J. (2011). Effects of nitrogen source and concentration on biomass and oil production of a Scenedesmus rubescens like microalga. *Bioresource Technology*, *102*(2), 1615-1621.

Liu, B., & Benning, C. (2013). Lipid metabolism in microalgae distinguishes itself. *Current opinion in biotechnology*, *24*(2), 300-309.

Liu, J., Mukherjee, J., Hawkes, J. J., & Wilkinson, S. J. (2013). Optimization of lipid production for algal biodiesel in nitrogen stressed cells of Dunaliella salina using FTIR analysis. *Journal of Chemical Technology & Biotechnology*, *88*(10), 1807-1814.

Liu, N., Guo, B., Cao, Y., Wang, H., Yang, S., Huo, H., ... & Niu, S. (2021). Effects of organic carbon sources on the biomass and lipid production by the novel microalga Micractinium reisseri FM1 under batch and fed-batch cultivation. *South African Journal of Botany*, *139*, 329-337.

Lyon, B. R., & Mock, T. (2014). Polar microalgae: new approaches towards understanding adaptations to an extreme and changing environment. *Biology*, *3*(1), 56-80.

Ma, C., Ren, H., Xing, D., Xie, G., Ren, N., & Liu, B. (2019). Mechanistic understanding towards the effective lipid production of a microalgal mutant strain Scenedesmus sp. Z-4 by the whole genome bioinformation. *Journal of hazardous materials*, *375*, 115-120.

Madeira, M. S., Cardoso, C., Lopes, P. A., Coelho, D., Afonso, C., Bandarra, N. M., & Prates, J. A. (2017). Microalgae as feed ingredients for livestock production and meat quality: A review. *Livestock Science*, *205*, 111-121.

Makareviciene, V., Andrulevičiūtė, V., Skorupskaitė, V., & Kasperovičienė, J. (2011). Cultivation of microalgae Chlorella sp. and Scenedesmus sp. as a potentional biofuel feedstock. *Environmental Research, Engineering and Management*, *57*(3), 21-27.

Maltsev, Y., & Maltseva, K. (2021). Fatty acids of microalgae: Diversity and applications. *Reviews in Environmental Science and Bio/Technology*, *20*, 515-547.

Maltsev, Y., Maltseva, K., Kulikovskiy, M., & Maltseva, S. (2021). Influence of light conditions on microalgae growth and content of lipids, carotenoids, and fatty acid composition. *Biology*, *10*(10), 1060.

Martins, C. F., Pestana, J. M., Alfaia, C. M., Costa, M., Ribeiro, D. M., Coelho, D., ... & Prates, J. A. (2021). Effects of Chlorella vulgaris as a feed ingredient on the quality and nutritional value of weaned piglets' meat. *Foods*, *10*(6), 1155.

Matos, Â. P. (2017). The impact of microalgae in food science and technology. *Journal of the American Oil Chemists' Society*, *94*, 1333-1350.

Mendoza, A., Morales, V., Sanchez-Bayo, A., Rodriguez-Escudero, R., Gonzalez-Fernandez, C., Bautista, L. F., & Vicente, G. (2020). The effect of the lipid extraction method used in biodiesel production on the integrated recovery of biodiesel and biogas from Nannochloropsis gaditana, Isochrysis galbana and Arthrospira platensis. *Biochemical Engineering Journal*, *154*, 107428.

Merchant, S. S., Prochnik, S. E., Vallon, O., Harris, E. H., Karpowicz, S. J., Witman, G. B., ... & Grossman, A. R. (2007). The Chlamydomonas genome reveals the evolution of key animal and plant functions. *Science*, *318*(5848), 245-250.

Minhas, A. K., Hodgson, P., Barrow, C. J., Sashidhar, B., & Adholeya, A. (2016). The isolation and identification of new microalgal strains producing oil and carotenoid simultaneously with biofuel potential. *Bioresource technology*, *211*, 556-565.

Misra, N., & Panda, P. K. (2013). In search of actionable targets for agrigenomics and microalgal biofuel production: sequence-structural diversity studies on algal and higher plants with a focus on GPAT protein. *OMICS: A Journal of Integrative Biology*, *17*(4), 173-186.

Mitra, M., Patidar, S. K., & Mishra, S. (2015). Integrated process of two stage cultivation of Nannochloropsis sp. for nutraceutically valuable eicosapentaenoic acid along with biodiesel. *Bioresource technology*, *193*, 363-369.

Molino, A., Iovine, A., Casella, P., Mehariya, S., Chianese, S., Cerbone, A., ... & Musmarra, D. (2018). Microalgae characterization for consolidated and new application in human food, animal feed and nutraceuticals. *International journal of environmental research and public health*, *15*(11), 2436.

Morançais, M., Mouget, J. L., & Dumay, J. (2018). Proteins and pigments. In *Microalgae in health and disease prevention* (pp. 145-175). Academic Press.

Morowvat, M. H., Rasoul-Amini, S., & Ghasemi, Y. (2010). Chlamydomonas as a "new" organism for biodiesel production. *Bioresource technology*, *101*(6), 2059-2062.

Muñoz, C. F., Südfeld, C., Naduthodi, M. I., Weusthuis, R. A., Barbosa, M. J., Wijffels, R. H., & D'Adamo, S. (2021). Genetic engineering of microalgae for enhanced lipid production. *Biotechnology advances*, *52*, 107836.

Mutanda, T., Naidoo, D., Bwapwa, J. K., & Anandraj, A. (2020). Biotechnological applications of microalgal oleaginous compounds: current trends on microalgal bioprocessing of products. *Frontiers in Energy Research*, *8*, 598803.

Nalley, J. O., O'Donnell, D. R., & Litchman, E. (2018). Temperature effects on growth rates and fatty acid content in freshwater algae and cyanobacteria. *Algal research*, *35*, 500-507.

Niu, Y. F., Wang, X., Hu, D. X., Balamurugan, S., Li, D. W., Yang, W. D., ... & Li, H. Y. (2016). Molecular characterization of a glycerol-3-phosphate acyltransferase reveals key features essential for triacylglycerol production in Phaeodactylum tricornutum. *Biotechnology for Biofuels*, *9*, 1-11.

Nzayisenga, J. C., Farge, X., Groll, S. L., & Sellstedt, A. (2020). Effects of light intensity on growth and lipid production in microalgae grown in wastewater. *Biotechnology for biofuels*, *13*, 1-8.

Ogbonna, I. O., Okpozu, O. O., Ikwebe, J., & Ogbonna, J. C. (2018). Utilisation of Desmodesmus subspicatus LC172266 for simultaneous remediation of cassava wastewater and accumulation of lipids for biodiesel production. *Biofuels*.

Okpozu, O. O., Ogbonna, I. O., Ikwebe, J., & Ogbonna, J. C. (2019). Phycoremediation of cassava wastewater by Desmodesmus armatus and the concomitant accumulation of lipids for biodiesel production. *Bioresource Technology Reports*, *7*, 100255.

Otero, P., Saha, S. K., Gushin, J. M., Moane, S., Barron, J., & Murray, P. (2017). Identification of optimum fatty acid extraction methods for two different microalgae Phaeodactylum tricornutum and Haematococcus pluvialis for food and biodiesel applications. *Analytical and Bioanalytical Chemistry*, *409*, 4659-4667.

Park, S., Kim, J., Park, Y., Kim, S., Cho, S., Yu, J., ... & Lee, T. (2018). Comparison of trophic modes to maximize biomass and lipid productivity of micractinium inermum NLP-F014. Biotechnology and Bioprocess Engineering, 23, 238-245.

Patel, A., Gami, B., Patel, P., & Patel, B. (2023). Biodiesel production from microalgae Dunaliella tertiolecta: A study on economic feasibility on large-scale cultivation systems. *Biomass Conversion and Biorefinery*, *13*(2), 1071-1085.

Patel, A., Rova, U., Christakopoulos, P., & Matsakas, L. (2020). Introduction to essential fatty acids. *Nutraceutical Fatty Acids from Oleaginous Microalgae: A Human Health Perspective*, 1-22.

Popovich, C. A., Damiani, C., Constenla, D., Martínez, A. M., Freije, H., Giovanardi, M., ... & Leonardi, P. I. (2012). Neochloris oleoabundans grown in enriched natural seawater for biodiesel feedstock: evaluation of its growth and biochemical composition. *Bioresource Technology*, *114*, 287-293.

Prathima, A., & Karthikeyan, S. (2017). Characteristics of micro-algal biofuel from Botryococcus braunii. *Energy Sources, Part A: Recovery, Utilization, and Environmental Effects*, *39*(2), 206-212.

Rahman, N. J. A., Ramli, A., Jumbri, K., & Uemura, Y. (2019). Tailoring the surface area and the acid–base properties of ZrO2 for biodiesel production from Nannochloropsis sp. *Scientific reports*, *9*(1), 16223.

Rengel, R., Smith, R. T., Haslam, R. P., Sayanova, O., Vila, M., & Leon, R. (2018). Overexpression of acetyl-CoA synthetase (ACS) enhances the biosynthesis of neutral lipids and starch in the green microalga Chlamydomonas reinhardtii. *Algal Research*, *31*, 183-193.

Rodríguez-Palacio, M. C., Cabrera-Cruz, R. B. E., Rolón-Aguilar, J. C., Tobías-Jaramillo, R., Martínez-Hernández, M., & Lozano-Ramírez, C. (2022). The cultivation of five microalgae species and their potential for biodiesel production. *Energy, Sustainability and Society*, *12*(1), 1-6.

Roopnarain, A., Gray, V. M., & Sym, S. D. (2014). Phosphorus limitation and starvation effects on cell growth and lipid accumulation in Isochrysis galbana U4 for biodiesel production. *Bioresource technology*, *156*, 408-411.

Saadaoui, I., Cherif, M., Siddiqui, S. A., El Anbari, M., Masmoudi, F., & Al Jabri, H. (2024). Abiotic stress applied to Tetraselmis subcordiformis reveals enhanced metabolic responses and increased production of Omega-3 fatty acids: A Metabolomic approach. *Algal Research*, *77*, 103338.

Sakarika, M., & Kornaros, M. (2019). Chlorella vulgaris as a green biofuel factory: comparison between biodiesel, biogas and combustible biomass production. *Bioresource technology*, *273*, 237-243.

Salama, E. S., Govindwar, S. P., Khandare, R. V., Roh, H. S., Jeon, B. H., & Li, X. (2019). Can omics approaches improve microalgal biofuels under abiotic stress?. *Trends in plant science*, *24*(7), 611-624.

Sánchez-Bayo, A., López-Chicharro, D., Morales, V., Espada, J. J., Puyol, D., Martínez, F., ... & Rodríguez, R. (2020). Biodiesel and biogas production from Isochrysis galbana using dry and wet lipid extraction: A biorefinery approach. *Renewable energy*, *146*, 188-195.

Saranya, G., & Ramachandra, T. V. (2021). Scope for biodiesel and bioactive compounds production in the diatom Nitzschia punctata. *Fuel, 300*, 120985.

Sarayloo, E., Simsek, S., Unlu, Y. S., Cevahir, G., Erkey, C., & Kavakli, I. H. (2018). Enhancement of the lipid productivity and fatty acid methyl ester profile of Chlorella vulgaris by two rounds of mutagenesis. *Bioresource technology, 250*, 764-769.

Sarayloo, E., Tardu, M., Unlu, Y. S., Simsek, S., Cevahir, G., Erkey, C., & Kavakli, I. H. (2017). Understanding lipid metabolism in high-lipid-producing Chlorella vulgaris mutants at the genome-wide level. *Algal research, 28*, 244-252.

Scheibel, S., de Oliveira, C. A. L., de Alvarenga Boyd, M., Pereira, L. C., Rodrigues, D. M., Barion, M. R. L., ... & Vasconcellos, R. S. (2021). DHA from microalgae Schizochytrium spp.(Thraustochytriaceae) modifies the inflammatory response and gonadal lipid profile in domestic cats. *British Journal of Nutrition, 126*(2), 172-182.

Schnorr, S. L., & Berry, D. (2022). Lipid synthesis at the trophic base as the source for energy management to build complex structures. *Current Opinion in Biotechnology, 73*, 364-373.

Schuhmann, H., Lim, D. K., & Schenk, P. M. (2012). Perspectives on metabolic engineering for increased lipid contents in microalgae. *Biofuels, 3*(1), 71-86.

Scranton, M. A., Ostrand, J. T., Fields, F. J., & Mayfield, S. P. (2015). Chlamydomonas as a model for biofuels and bio-products production. *The Plant Journal, 82*(3), 523-531.

Shang, C. H., Zhu, S. N., Yuan, Z. H., & Wang, Z. M. (2012). Molecular cloning and characterization analysis of malic enzyme gene from Dunaliella parva. *Advanced Materials Research, 347*, 2536-2540.

Sharma, T., Gour, R. S., Kant, A., & Chauhan, R. S. (2015). Lipid content in Scenedesmus species correlates with multiple genes of fatty acid and triacylglycerol biosynthetic pathways. *Algal research, 12*, 341-349.

Sheikhzadeh, N., Soltani, M., Heidarieh, M., & Ghorbani, M. (2024). Role of Dietary Microalgae on Fish Health and Fillet Quality: Recent Insights and Future Prospects. *Fishes, 9*(1), 26.

Shi, L., & Tu, B. P. (2015). Acetyl-CoA and the regulation of metabolism: mechanisms and consequences. *Current opinion in cell biology, 33*, 125-131.

Shi, T. Q., Wang, L. R., Zhang, Z. X., Sun, X. M., & Huang, H. (2020). Stresses as first-line tools for enhancing lipid and carotenoid production in microalgae. *Frontiers in bioengineering and biotechnology, 8*, 610.

Shtaida, N., Khozin-Goldberg, I., & Boussiba, S. (2015). The role of pyruvate hub enzymes in supplying carbon precursors for fatty acid synthesis in photosynthetic microalgae. *Photosynthesis research, 125*, 407-422.

Silitonga, A. S., Masjuki, H. H., Ong, H. C., Mahlia, T. M. I., & Kusumo, F. (2017). Optimization of extraction of lipid from Isochrysis galbana microalgae species for biodiesel synthesis. *Energy Sources, Part A: Recovery, Utilization, and Environmental Effects, 39*(11), 1167-1175.

Singh, P., Kumari, S., Guldhe, A., Misra, R., Rawat, I., & Bux, F. (2016). Trends and novel strategies for enhancing lipid accumulation and quality in microalgae. *Renewable and Sustainable Energy Reviews, 55*, 1-16.

Singh, P., Kumari, S., Guldhe, A., Singh, G., & Bux, F. (2017). ACCase and rbcL gene expression as a function of nutrient and metal stress for enhancing lipid productivity in Chlorella sorokiniana. *Energy Conversion and Management, 148*, 809-819.

Spolaore, P., Joannis-Cassan, C., Duran, E., & Isambert, A. (2006). Commercial applications of microalgae. *Journal of bioscience and bioengineering, 101*(2), 87-96.

Suganya, T., Varman, M., Masjuki, H. H., & Renganathan, S. (2016). Macroalgae and microalgae as a potential source for commercial applications along with biofuels production: A biorefinery approach. *Renewable and Sustainable Energy Reviews, 55*, 909-941.

Suh, H. S., Do, J. M., Yeo, H. T., & Yoon, H. S. (2024). Cattle wastewater treatment using green microalga Coelastrella sp. KNUA068 as a promising bioenergy feedstock with enhanced biodiesel quality. *Water Science & Technology, 89*(3), 714-729.

Suzuki, K. (2017). Large-scale cultivation of Euglena. *Euglena: biochemistry, cell and molecular biology*, 285-293.

Tang, H., Abunasser, N., Garcia, M. E. D., Chen, M., Ng, K. S., & Salley, S. O. (2011). Potential of microalgae oil from Dunaliella tertiolecta as a feedstock for biodiesel. *Applied Energy, 88*(10), 3324-3330.

Tasić, M. B., Pinto, L. F. R., Klein, B. C., Veljković, V. B., & Maciel Filho, R. (2016). Botryococcus braunii for biodiesel production. *Renewable and sustainable energy reviews, 64*, 260-270.

Tizvir, A., Shojaeefard, M. H., Zahedi, A., & Molaeimanesh, G. R. (2022). Performance and emission characteristics of biodiesel fuel from Dunaliella tertiolecta microalgae. *Renewable Energy, 182*, 552-561.

Torres-Tiji, Y., Fields, F. J., & Mayfield, S. P. (2020). Microalgae as a future food source. *Biotechnology advances, 41*, 107536.

Toyoshima, M., & Sato, N. (2015). High-level accumulation of triacylglycerol and starch in photoautotrophically grown Chlamydomonas debaryana NIES-2212. *Plant and Cell Physiology, 56*(12), 2447-2456.

Tran, N. A. T., Padula, M. P., Evenhuis, C. R., Commault, A. S., Ralph, P. J., & Tamburic, B. (2016). Proteomic and biophysical analyses reveal a metabolic shift in nitrogen deprived Nannochloropsis oculata. *Algal Research, 19*, 1-11.

Úbeda-Mínguez, P., García-Maroto, F., & Alonso, D. L. (2017). Heterologous expression of DGAT genes in the marine microalga Tetraselmis chui leads to an increase in TAG content. *Journal of Applied Phycology, 29*, 1913-1926.

Vadivel, M., Arvindnarayan, S., Kumar, G., Shobana, S., Dharmaraja, J., Nguyen, D. D., ... & Prabhu, K. K. S. (2020). Biodiesel potentiality of microalgae species: evaluation using various nitrogen sources. *Waste and Biomass Valorization, 11*, 1671-1679.

Valledor, L., Furuhashi, T., Recuenco-Muñoz, L., Wienkoop, S., & Weckwerth, W. (2014). System-level network analysis of nitrogen starvation and recovery in Chlamydomonas reinhardtii reveals potential new targets for increased lipid accumulation. *Biotechnology for biofuels, 7*(1), 1-17.

Vélez-Landa, L., Hernández-De León, H. R., Pérez-Luna, Y. D. C., Velázquez-Trujillo, S., Moreira-Acosta, J., Berrones-Hernández, R., & Sánchez-Roque, Y. (2021). Influence of light intensity and photoperiod on the photoautotrophic growth and lipid content of the microalgae Verrucodesmus verrucosus in a photobioreactor. *Sustainability, 13*(12), 6606.

Wagner, M., Hoppe, K., Czabany, T., Heilmann, M., Daum, G., Feussner, I., & Fulda, M. (2010). Identification and characterization of an acyl-CoA: diacylglycerol acyltransferase 2 (DGAT2) gene from the microalga O. tauri. *Plant Physiology and Biochemistry, 48*(6), 407-416.

Wei, H., Shi, Y., Ma, X., Pan, Y., Hu, H., Li, Y., ... & Liu, J. (2017). A type-I diacylglycerol acyltransferase modulates triacylglycerol biosynthesis and fatty acid composition in the oleaginous microalga, Nannochloropsis oceanica. *Biotechnology for Biofuels, 10*, 1-18.

Wu, S., Feng, B., Li, K., Zhu, X., Liang, S., Liu, X., ... & Fan, D. (2012). Fish consumption and colorectal cancer risk in humans: a systematic review and meta-analysis. *The American journal of medicine, 125*(6), 551-559.

Wu, T., Mao, X., Kou, Y., Li, Y., Sun, H., He, Y., & Chen, F. (2019). Characterization of microalgal acetyl-CoA synthetases with high catalytic efficiency reveals their regulatory mechanism and lipid engineering potential. *Journal of agricultural and food chemistry, 67*(34), 9569-9578.

Xu, J., Zheng, Z., & Zou, J. (2009). A membrane-bound glycerol-3-phosphate acyltransferase from Thalassiosira pseudonana regulates acyl composition of glycerolipids. *Botany, 87*(6), 544-551.

Xue, J., Niu, Y. F., Huang, T., Yang, W. D., Liu, J. S., & Li, H. Y. (2015). Genetic improvement of the microalga Phaeodactylum tricornutum for boosting neutral lipid accumulation. *Metabolic engineering, 27*, 1-9.

Yamaoka, Y., Achard, D., Jang, S., Legéret, B., Kamisuki, S., Ko, D., ... & Lee, Y. (2016). Identification of a Chlamydomonas plastidial 2-lysophosphatidic acid acyltransferase and its use to engineer microalgae with increased oil content. *Plant biotechnology journal, 14*(11), 2158-2167.

Yang, M., Zhou, M., & Song, L. (2020). A review of fatty acids influencing skin condition. *Journal of Cosmetic Dermatology, 19*(12), 3199-3204.

Zhang, Y., Gu, Z., Ren, Y., Wang, L., Zhang, J., Liang, C., ... & Ye, N. (2021). Integrating transcriptomics and metabolomics to characterize metabolic regulation to elevated CO 2 in Chlamydomonas Reinhardtii. *Marine Biotechnology, 23*, 255-275.

Zhao, J., Ge, Y., Liu, K., Yamaoka, Y., Zhang, D., Chi, Z., ... & Kong, F. (2023). Overexpression of a MYB1 transcription factor enhances triacylglycerol and starch accumulation and biomass production in the green microalga Chlamydomonas reinhardtii. *Journal of Agricultural and Food Chemistry, 71*(46), 17833-17841.

Zheng, J. S., Hu, X. J., Zhao, Y. M., Yang, J., & Li, D. (2013). Intake of fish and marine n-3 polyunsaturated fatty acids and risk of breast cancer: meta-analysis of data from 21 independent prospective cohort studies. *Bmj, 346*, f3706.

Zienkiewicz, K., Du, Z. Y., Ma, W., Vollheyde, K., & Benning, C. (2016). Stress-induced neutral lipid biosynthesis in microalgae - molecular, cellular and physiological insights. *Biochimica et Biophysica Acta (BBA)-Molecular and Cell Biology of Lipids, 1861*(9), 1269-1281.

Zulu, N. N., Zienkiewicz, K., Vollheyde, K., & Feussner, I. (2018). Current trends to comprehend lipid metabolism in diatoms. *Progress in lipid research, 70*, 1-16.

Chapter 6

Uncovering the Botanical Alarms: Electrical Signaling in Response to Insect Herbivory - A Review

Gabriela Niemeyer Reissig[1,*], PhD
Thiago Francisco de Carvalho Oliveira[1], MS
Douglas Antônio Posso[1], PhD
Helena Chaves Tasca[1], MS,
Luis Felipe Basso[1], B.Sc
Lyana Pintos Ramos[1], B.Sc
João Gabriel Moreira de Souza[1], B.Sc
André Geremia Parise[2], MS
and Dori Edson Nava[3], PhD

[1]Laboratory of Plant Cognition and Electrophysiology, Department of Botany, Institute of Biology, Federal University of Pelotas, Pelotas, Brazil
[2]School of Biological Sciences, University of Reading, Reading, UK
[3]Laboratory of Entomology, Embrapa Clima Temperado, Pelotas, Brazil

Abstract

Plant electrophysiology has emerged as an important field within plant physiology, offering a means to detect responses to various stresses before the manifestation of classic symptoms. Among these stresses, insect predation stands out for its ability to damage plant tissues and cells,

---

* Corresponding Author's Email: gabriela.niemeyer.reissig@gmail.com.

In: Advances in Biology. Volume 6
Editor: Charles D. Grant
ISBN: 979-8-89113-712-7

triggering defense mechanisms such as the induction of systemic responses via jasmonates, oxidative stress, and the generation of electrical signals. Research focusing on herbivory and electrical signals enhances our understanding of plant perception and response mechanisms, shedding light on how information propagates within plants. This study aims to provide a comprehensive review of electrical signals associated with herbivory in plants, encompassing ecological perspectives, signal types, generation mechanisms, detection, analysis techniques, and future applications. The most observed electrical signal during herbivory is the variation potential (VP), also known as slow wave potentials (SWP), documented in plants attacked by various insects. Electrical signals originate at the site of injury and propagate to systemic tissues, triggering responses in different parts. Molecules such as glutamate, ion channels like GLR3, plasma membrane H^+-ATPase such as AHA1, and other signals like hydraulic mechanisms are involved in herbivory-induced electrical signaling. Beyond individual signals, exploring the plant electrome is crucial for understanding plant defense mechanisms. The study of plant electrome involves measuring the bioelectrical activity of plants as changes in the electrical potential of tissues over time. Investigations into caterpillar predation on tomato fruits and wheat infested with pathogenic fungi illustrate diverse applications of the electrome in understanding plant electrical signaling and allowing early disease diagnosis. Various techniques, including electrography, intracellular microelectrodes, and patch clamp, have been developed to accurately capture and analyze electrophysiological data. Despite challenges such as spatial limitations and technical complexity, these techniques significantly advance our understanding of plant electrical dynamics, benefiting agriculture and biotechnology. Integrating these approaches with advanced data acquisition techniques holds potential for unraveling intricate plant responses, particularly in the context of insect predation. Understanding plant electrical dynamics under insect herbivory can inform crop management strategies, enabling non-destructive monitoring and aiding in precise chemical and biological controls essential for sustainable agriculture amidst the need to minimize agrochemical use.

Keywords: electrome, long-distance signaling, plant defense, plant-insect interaction, plant electrophysiology

1. Introduction

Herbivory is a form of predatory ecological interaction in which herbivores feed on plants or parts thereof. It influences the structure and composition of plant communities, the distribution of nutrients in ecosystems, and the interactions among different species of plants and animals (Shatz et al., 2017). Also, it poses a significant challenge to agriculture, affecting crop yield and quality worldwide. Herbivore insects consume plant tissues, leading to direct damage such as defoliation, stem boring, and fruit destruction, as well as indirect damage through the transmission of plant diseases (Culliney, 2014).

Wound caused by insect predation can trigger a range of responses in plants, including defense mechanisms such as the production of toxic chemical compounds and the release of chemical signals to attract herbivore predators (Aartsma et al., 2017). To enact these responses, plants necessitate mechanisms for swift reaction, which rely on a complex internal signaling network integrating chemical, hydraulic, and electrical signals. This network enables plants to alert other parts of their bodies not yet attacked by herbivores, generating a systemic response, or even neighboring plants, about the attack (Choi et al., 2017).

In studies that analyze the electrical activity of single cells or groups of cells, the most observed electrical signals during herbivory are variation potentials (VPs), also known as slow wave potentials (SWPs). Several studies have documented this type of signal in plants subjected to predation by various insects (Zimmermann et al., 2009; Mousavi et al., 2013; Zimmermann et al., 2016; Pachú et al., 2021). In addition to individual signals, the study of plant electrome has proven to be a valuable way to understand electrical signaling. The concept of plant electrome is based on the general definition of "electrome" by De Loof (2016), which refers to the totality electrical activity (ionic currents and electron flow), of any living entity, from the cell to the whole organism level. Thus, plant electrome corresponds to the bioelectrical activity of plants measured as micro-voltage changes in stimulated or non-stimulated tissues (Saraiva et al., 2017; Toledo et al., 2019). Studies involving caterpillar predation on tomato fruits (Reissig et al., 2021) and wheat infested with pathogenic fungi (Simmi et al., 2023) demonstrate different applications of the electrome analysis, be it for understanding electrical signaling in plants or as a tool for early disease diagnosis, respectively.

Several techniques for capturing and analyzing electrophysiological data, such as electrography, intracellular microelectrodes, and patch clamp, have been developed to accurately capture electrical signals. However, each technique presents its own challenges, ranging from spatial limitations to technical complexity. Despite these challenges, the application of these techniques has significantly contributed to our understanding of plant electrical dynamics, with implications for agriculture and biotechnology. Moving forward, novel analytical methods like Time Dispersion Analysis of Features (TDAF) open the possibility to study the temporal complexity of electrophysiological data, providing a more comprehensive understanding of the electrome (Souza et al., 2017; Costa et al., 2023). Integrating such approaches with advanced data acquisition techniques holds potential for unraveling the intricate electrical responses of plants, particularly in the context of insect predation (Reissig et al., 2021).

Understanding the electrical dynamics of plants under insect herbivory can assist in the management of different crops. The damages caused lead to alterations in plant structure and physiology, which can be identified through electrography techniques and classified using, for example, Machine Learning techniques. Ultimately, this knowledge can facilitate more sustainable crop management, enabling non-destructive monitoring and aiding in chemical and biological controls, such as the more precise application of insecticides and fungicides. This is particularly crucial given the increasing necessity for sustainable techniques that minimize the use of agrochemicals.

The aim of this chapter is to provide a comprehensive review of the electrical signals associated with herbivory in plants. It encompasses the ecological perspective of plant-insect interaction, the types of signals present in plants when attacked by insects and their associated responses, as well as mechanisms for generating and propagating these signals and techniques for their detection and analysis. Additionally, the review concludes by discussing future perspectives on the application of this knowledge.

2. Plant-Insect Interaction

Plants and insects have developed survival mechanisms over evolutionary time through interactions between them. These interactions have their roots millions of years ago, with vestige of notable coevolution due to pollination, for example, found since the late Cretaceous period, about 75 million years ago (Cruaud et al., 2012; Schatz et al., 2017), and traces of herbivory found much earlier, through fossils of the mouthparts of phytophagous insects between the late Silurian and early Devonian periods, around 420 million years ago (Labandeira, 2006; Schatz et al., 2017). Moreover, there are traces from the Middle Triassic period, about 240 million years ago, indicating that mites, orthopteroids, hemipteroids, and basal clades of Holometabola were already colonizing ferns and gymnosperms (Labandeira, 2006; Schatz et al., 2017). Thus, some insects may have developed specific characteristics to feed on certain parts of plants or co-evolved with them, which may have implications for both parties. During the Cretaceous period, angiosperms experienced rapid expansion and diversification, significantly influencing the evolution of insects, as many insect groups coevolved with angiosperms, giving rise to a true plant-related fauna (Labandeira, 2006; Schatz et al., 2017), exhibiting various types of mutualistic interactions such as the classic pollination mutualism, among other examples where both groups benefit. Therefore, although herbivory is an ancient problem for plants, some insects have coevolved with them to benefit both parties.

The most well-known mutualistic interaction between plants and insects is pollination; however, there are other types of mutualistic interactions between these organisms, where insects can benefit from food provided by plants, and plants, in turn, receive protection from other herbivores. An example of this type of mutualism is found in myrmecotrophy, between the plant *Cecropia obtusa* and the ants *Azteca ovaticeps* or *Azteca alfari*, where the ants defend their host trees from enemies that, in turn, provide shelter (domatia) and food (extrafloral nectar, Müllerian bodies) to them (Dejean et al., 2012). In addition to mutualistic interactions, there is commensalism, where only one of the involved parts benefits. In this context, there is the example of sexual deception, famously in orchids. For example, the orchid *Drakae glyptodon*, which can mimic the shape and scent of female wasps of the species *Zaspilothynnus trilobatus*, attracts males that copulate with the flowers. The delusional males end up covered in pollen without the desired reward (Gaskett, 2011).

As previously described, herbivory has been a problem for plants for thousands of years, and over this period of evolution, plants have developed sophisticated defense mechanisms to combat attacks from insects adapted to feeding on their parts, whether by recruiting ants to defend them in exchange for shelter and food or by recruiting herbivore predators to attack them. In the latter example lies the well-known tritrophic relationship, where plants are attacked by an herbivore predator such as caterpillars or aphids emit chemical substances in the air, called volatile organic compounds (VOCs), to attract other insects, such as predatory wasps, which lay their eggs on them or kill the herbivores, thus disinfesting the plant (Aartsma et al., 2017).

There is a range of attraction and repellence mechanisms involved in defense against herbivores, just as herbivory interactions also trigger adaptive resistance/immunity responses in plants. These interactions and responses are possible due to the plants' ability to interpret their environment, including the recognition of the identity of the beings around them, in this case, insects (Baluska and Mancuso, 2021). Unlike interactions with large animals that could consume the plant as a whole, in the plant-insect interaction, the plant develops immunity and specific attraction and repulsion mechanisms for insects. In positive interactions, attractive mechanisms have been developed, where plants emit compounds such as VOCs into the air and root exudates into the soil to attract desired insects, or repel undesired ones (Zhou and Jander, 2022; Grunseich et al., 2020). Conversely, due to negative interactions such as herbivory, defense mechanisms have been developed to repel these insects and prevent some from settling on the plant, producing defense molecules such as VOCs, as well as immune responses. Many of these defense substances are known, such as methyl jasmonate (MeJA) and methyl salicylate (MeSA).

Plant defense and immunity are divided in two main parts: (i) morphological barriers such as thorns, trichomes, and the plant cuticle; and (ii) chemical barriers such as the salicylic acid pathway, jasmonic acid pathway, ethylene, hormones such as auxins, gibberellins, cytokinins, and brassinosteroids (Lazebnik et al., 2014; Mithöfer and Boland, 2012; Núñez-Farfán, Fornoni, & Valverde, 2007), herbivore-induced plant volatiles (HIPVs) that can repel herbivores or attract their predators (Thompson and Goggin, 2006; Heil and Silva Bueno, 2007), and other chemical compounds that can increase mortality and hinder reproduction (Karabourniotis et al., 2020; Kortbeek et al., 2021; Pieterse et al., 2014; Haas et al., 2018). When observing this plant relationship in defense against harmful organisms, it is termed induced resistance, characterized by the plant's defense or immune response having a reducing effect on the damage caused by organisms or their population, thus also demonstrating a decrease in the hostile activity of such organisms (Dean and Kuć 1985; Schultz 1988; Kloepper et al., 1992; Van Loon 2000; Kuć 2001), whether phytophagous insects or herbivory-related pathogens (De Bruxelles and Roberts, 2001; Moran and Thompson, 2001; Liu et al., 2020; Lee et al., 2012), with induced resistance also dependent on the plant's pre-existing components, which will determine both the activation of defense-related genes and the synthesis of proteins and metabolic changes related to defense (Kloepper et al., 1992; Van Loon 2000; Kuć 2001; Karban and Baldwin, 2007).

Induced resistance can also be direct or indirect, with direct resistance related to the production of insecticidal compounds or compounds that hinder herbivory (Karban and Baldwin, 2007), often associated with pathways such as salicylic acid, jasmonic acid, and ethylene, while indirect resistance is related to insect dispersal or attraction of their predators (Dicke et al., 2003) through HIPVs, with many being terpenes (Ashour et al., 2010). Much of these induced defense responses are due to the recognition of compounds derived from herbivores or pathogens and thus defending themselves according to the identified threat. Considering plant immunity, the main means are the jasmonic acid, salicylic acid, and ethylene pathways (Er et al., 2012; Howe and Jander, 2008); these pathways are activated by diverse predator activities but can have positive effects on each other. The jasmonic acid pathway is influenced by chewing insects and necrotrophic pathogens (Thaler et al., 2012; Smith and Clement, 2012), the salicylic acid pathway is activated by piercing-sucking insects and biotrophic pathogens (Walling, 2000), and the ethylene pathway by both insects (Smith and Clement, 2012; Zhu et al., 2018; Qi et al., 2020). However, recent studies have shown that both the jasmonic acid and salicylic acid pathways can be activated by different plant resistance inducers (Martinez de Ilarduya et al., 2003; Rubil et al., 2022; Züst and Agrawal, 2016; De Vos and Jander, 2010; Bonaventure, 2012).

In comparison to the immune systems of other species, plant immunity operates similarly, identifying and recognizing common patterns. When a plant is attacked by pathogens, it utilizes pathogen-associated molecular patterns (PAMPs) and herbivore-associated molecular patterns (HAMPs), which are recognized by pattern recognition receptors (PRRs) which occur in the cells membranes as receptor-like kinases or as protein with receptor-like nature (Acevedo et al., 2015; Stahl et al., 2015; Saijo et al., 2018). Recognition can also occur following tissue damage in the plant. This is called damage-associated molecular patterns (DAMPs), which represent the plant's immune system identifying damage to its own tissue and discerning where and why it occurred, thus addressing the source as well. DAMP recognition also occurs through stimulation of the cell membrane, resulting in the accumulation of $[Ca^{2+}]_{cyt}$, which can be triggered by herbivore or pathogen wounds (Maffei et al., 2004; Schäfer et al., 2011; Yan et al., 2018). These stimuli and changes in ion dynamics support the generation and propagation of electrical signals, which play a role in plant immunity. Electrical signals are considered prominent in activating the immune system as they are precursors to immune chemical signaling and pathways responsible for plant immunity (Maffei, Mithofer, Boland 2007; Zebelo, Maffei, 2015). Next, we will explore plant electrical signals associated with herbivory, addressing the existing and most prominent types, as well as examples of studies demonstrating the identification of these signals in the interaction between different insects and plants.

3. Herbivory and Electrical Signals

In the intricate realm of plant biology, electrical signaling stands out as a fundamental aspect of the electrome, shaping the dynamic responses to various stimuli, including herbivory. Plants function as living circuits, relying on electrical activity for their proper functioning (Volkov et al., 2009; Stavrinidou et al., 2015; Toledo et al., 2019). A crucial aspect of this electrical activity is electrical signaling, involving ionic imbalances that propagate signals to inform nearby or distant parts of the plant (Baluška and Mancuso, 2013a; Vodeneev et al., 2016a, b). The entirety of a plant's electrical activity, including signaling, corresponds to the electrical dimension of plant life, and is known as the "electrome" in the context of omics sciences (de Loof, 2016; Souza et al., 2017; de Toledo et al., 2019).

The roots of plant electrical signaling are at the groundbreaking work of early pioneers of the late 19th and early 20th century, including Sir Jagadish Chandra Bose and Sir John Scott Burdon-Sanderson. Initially perceived as exclusive to sensitive plants (plants that present rapid leaf movement), subsequent investigations have expanded the recognition of action potentials (APs) and other electrical signals to a diverse array of plant species (Shepherd, 2005, 2012). Umrath's 1930 experiments marked a milestone, providing the first intracellular measurements of an action potential in vegetal cells (Volkov, 2006; Fromm and Lautner, 2007).

Contemporary research has unveiled striking parallels in voltage variations between animal and plant cells, shedding light on shared ionic channels and pumps encoded by homolog genes (Trewavas, 2016, 2017; Canales et al., 2018). Neurotransmitters, such as glutamate, traditionally associated with animal nerves, have emerged as pivotal components in plant signaling transmission and integration (Baluška et al., 2006; Brenner et al., 2007).

These electrical signals are results of the intricated relation on the plant's sessile lifestyle, as environmental fluctuations induce continuous electrical signals throughout plant tissues, categorized so far in four main types: action potentials (APs), variation potentials (VPs) that are also known as slow wave potentials (SWPs), local electric potentials (LEPs), and systemic potentials (SPs) (Fromm and Lautner, 2007; Huber and Bauerle, 2016; Sukhova et al., 2017; Szechynska-Hebda et al., 2017).

APs, characterized by rapid depolarization and swift repolarization, are triggered by stimuli such as electrical stimulation, thermal shock, or herbivore-induced damage (Vodeneev et al., 2016a; Sukhova et al., 2017). Calcium entry induces depolarization, activating Ca^{2+} channels, while Cl^- channels open to facilitate repolarization through K^+ efflux. H^+ pumps contribute to re-establishing the resting potential (Fromm and Lautner, 2007; Huber and Bauerle, 2016; Sukhova et al., 2017; Trebacz et al., 2006). APs propagate through plasmodesmata in phloem parenchymal cells, reaching speeds up to 3000 cm s^{-1}, resembling animal neurons (Volkov et al., 2000; Fromm and Lautner, 2007; de Loof, 2016).

VPs/SWPs exhibit rapid depolarization and slower repolarization, induced by various biotic and abiotic stimuli (Huber and Bauerle, 2016; Vodeneev et al., 2016b). Mechanical wounds, herbivore attacks, or environmental changes trigger VPs. The depolarization mechanism mirrors that of APs, while repolarization involves a transitory inactivation of H^+ pumps (Stahlberg et al., 2006a,b; Vodeneev et al., 2016b). VPs propagate in the xylem, diverging from the all-or-none principle observed in APs, with velocity and amplitude decreasing with distance from the stimulus (Stahlberg et al., 2006a,b; Vodeneev et al., 2016a).

Initiated by injury and stimulation with inorganic cations, SPs are characterized by apoplast hyperpolarization and do not obey the all-or-none principle (Zimmermann et al., 2009; Zimmermann et al., 2016). Activated by H^+ pumps, SPs propagate at speeds of 5–10 cm min^{-1}, with amplitude varying based on stimulus intensity. Unlike APs, SPs can be triggered by stimuli insufficient for AP initiation, offering a broader range of responses (Zimmermann et al., 2009; Zimmermann et al., 2016).

Generated at the stimulus site under diverse influences, such as light, temperature changes, water, nutrients, humidity, phytohormones, and pathogen infections, LEPs exhibit amplitude dependence on stimulus intensity and duration. They arise from the transient inactivation of H^+-ATPases and changes in ionic channel activity (Yan et al., 2009; Vodeneev et al., 2016a). LEPs contribute to nuanced and localized responses to various environmental stimuli.

Herbivory and mechanical wounding in plants have been shown to elicit electrical signals—mediated by two glutamate receptor-like proteins—that induce defense responses at local and distant sites. Mousavi et al., (2013) used

Arabidopsis thaliana plants and Egyptian cotton leafworm (*Spodoptera littoralis*) larvae as a model for investigating plant-herbivore interactions. The researchers positioned larvae on individual leaves, utilizing electrodes grounded in the soil and on leaf surfaces to monitor changes in electrical potentials. Notably, the leaf-surface potential remained unaltered as larvae traversed a leaf, but once feeding commenced, electrical signals emerged near the site of attack, subsequently spreading to neighboring leaves at a maximum speed of 9 centimeters per minute.

As reported by Mousavi et al. (2013), at locations receiving the electrical signals, jasmonate-mediated gene expression was activated, initiating the expression of defense-responsive genes. In a mutant *A. thaliana* plant lacking the jasmonate receptor, electrical signals were propagated, yet no defense response was elicited. Notably, defense responses were absent at remote sites when the transmission of the electrical signal was impeded by ablating the damaged leaf before the signal traversed the leaf petiole. These intriguing findings underscore the pivotal role of electrical signal generation and propagation in initiating defense responses at distant sites following herbivore attacks.

Relating membrane potential to hormonal synthesis, Marhav et al. (2019) study, conducted with electrodes on the root epidermis, demonstrates that nematode attack or injury to a single cell in *A. thaliana* roots triggers membrane depolarization and ethylene signaling.

Despite the conventional understanding that herbivore salivary secretions contain elicitor molecules recognized by host plants, inducing jasmonate-mediated defense responses, Mousavi and colleagues (2013) discovered that extensive mechanical wounding in the absence of herbivory also initiated electrical signal transmission and jasmonate biosynthesis. Furthermore, they demonstrated that applying electric pulses to mimic plant signals could artificially induce an herbivore-response gene-expression pattern. The study underscores the lingering uncertainty regarding how electrical signals are interpreted to stimulate jasmonate biosynthesis (Mousavi et al., 2013).

Despite herbivory in agricultural settings often resulting from insect attacks, herbivory caused by small arachnids, such as mites, also occurs. By studying the general electrome of tomato plants using sensors placed near the damaged leaf and on the stem and analyzing the electrome with Machine Learning, Najdenovska et al. (2021) indicated that it is possible to utilize electrophysiology to study tomato plant responses to mite infestation. Despite being a preliminary study, clear differences in the electrical activity of tomato plants in the presence of mites exist, aiming to understand and enhance initial agricultural control practices, thereby preventing large-scale infestations.

In another study involving tomatoes, specifically examining the electrical signal profile of green and ripe fruits attacked by *Helicoverpa armigera* caterpillars, Reissig et al. (2021) found that contrary to the expectation that fruits are merely plant organs capable of receiving information from the parent plant, they actually engage in bidirectional communication. In fact, they actively communicate in both directions, primarily alerting other modules (leaves) of the aerial part when under predation, through electrical signals often accompanied by a production of ROS in distal parts of the plant.

In a study inducing artificial injury, Costa et al. (2023) observed alterations in electrical signaling among different plant modules, whether the one subjected to the injury or distant modules. Thus, the electrome of bean plants underwent changes, and the observed electrical signal was accompanied by biochemical and hydraulic responses. This makes it possible to distinguish that not only substances present in insect saliva are capable of inducing electrical signals and biochemical responses, but also mechanical injuries.

Zimmermann et al. (2016), in an investigation into herbivore-triggered electrophysiological responses (elRs) across various plant and insect species, has highlighted a consistent ability of feeding herbivores to induce elRs, both locally and systemically. While the existence of various types of elRs in response to herbivore feeding is a common observation, VPs are limited to detection near the wounded plant area due to their inability to self-propagate. The long-distance transmission of APs relies on electrophysiological connectivity among individual plant cells, which seems constrained within plant tissue, leading to a loss of APs during their journey through the plant body. Intriguingly, their findings reveal that the subsequent hyperpolarization (related to SPs) following depolarizing events of the plasma membrane is capable of self-propagation, providing a plausible explanation for its high detectability in systemic plant parts. Future studies are warranted to decipher the potential information content of SPs; however, the observations suggest a natural relevance of SPs, especially in response to herbivore feeding as a natural stimulus. The studied plant species, including *Vicia faba*, *Hordeum vulgare*, *Nicotiana tabacum*, *Brassica napus*, and *Cucurbita maxima* interactions with the larvae of different lepidoptera species plus an aphid species facilitated a comprehensive understanding of herbivore-induced electrophysiological responses, paving the way for future investigations into the nuanced interplay between herbivore feeding, electrical signaling, and plant physiology.

Decoding the complex propagation of electrical signals in plants involves distinct pathways. Locally, within the parenchyma, signals spread cell-to-cell via plasmodesmata, while systemic spreading faces challenges due to the resistance posed by the small diameter of plasmodesmata pores (Zimmermann et al., 2009; Hedrich et al., 2016; Vodeneev et al., 2016a). The phloem network emerges as a primary pathway for APs and SPs, with sieve tube elements facilitating rapid transport of solutes and signaling molecules, including electrical signals. APs and SPs traveling along the sieve tube elements can excite adjacent parenchymal cells via plasmodesmata, ensuring long-distance communication (Vodeneev et al., 2016a; Baluska and Mancuso, 2009a, b; Hedrich et al., 2016). Following, we will address the current understanding of the generation and propagation of electrical signals in plants under herbivory, discussing the existing models and the involved components as well.

4. Generation and Propagation of Electrical Signals in Plants under Herbivory

Wound damage resulting from insect herbivory induces local and systemic responses in plants. At the local tissue level, where the damage occurred, some responses to the injury include wound healing and tissue regeneration (Lee and Seo, 2022). In tissues distant from the site of injury, the systemic defense response involves the activation of genes (Choi et al., 2017) and the antioxidant system (Reissig et al., 2021), as well as the production of jasmonate (Hu et al., 2022), among other factors. Systemic responses depend on a well-coordinated long-distance signaling. To achieve this, it is understood that the integration of chemical, hydraulic, and electrical signals is essential for plants to communicate internally, utilizing information of different natures to gain a better understanding of the overall picture of what is happening. Studies demonstrate that a network composed of a trio of messengers interacting with each other—reactive oxygen species (ROS), Ca^{2+}, and electrical signaling—is involved in long-distance signaling to various environmental disturbance sources, including insect predation (Choi et al., 2017). Farmer et al. (2020) proposed a model for wound-response leaf-to-leaf electrical signaling, based on rapid electro-osmotic coupling between the phloem and xylem. The models and underlying molecular components of the studies mentioned here will be further detailed in this section.

The transmission of electrical signals in plants attacked by insects depends on glutamate receptor-like (GLR) proteins, which are a family of ion channels permeable to cations and act in plant defense against biotic stresses (Toyota et al., 2018; Yan et al., 2024). In an elegant study using *A. thaliana* plants expressing the GCaMP3 fluorescent protein-based $[Ca^{2+}]_{cyt}$ sensor, it was observed that shortly after caterpillar attack, the cytosolic calcium concentration at the site of damage increased within 2 seconds, and it took 1 to 2 minutes to reach distant leaves (Toyota et al., 2018). Through studies with unique mutants *glr3.3* and *glr3.6*, the authors observed that for the rapid propagation of systemic electrical signals, the *GLR3.3* and *GLR3.6* ion channels were important, influencing the kinetics of cytosolic calcium propagation. The amino acid glutamate (Glu) is crucial for regulating these channels. Its increase for the activation of GLR ion channels can occur through damaged cells or actively released when plants undergo injury. In experiments conducted in the same study, the authors noticed that applying less L-glutamate restricted the extent of the increase in systemic cytosolic calcium concentration.

The AHA1 (Activator of Hsp90 ATPase 1) is an essential component for effectively repolarizing the membrane after injury. Electrogenic proton pumps, such as AHA1, are associated with the generation and duration of Slow Wave Potentials (SWPs) and control the wound response jasmonate pathway activation (Kumari et al., 2019; Farmer et al., 2020). In studies conducted with Arabidopsis, using *aha1-7* mutants and aphid electrodes for capturing electrical signals, the authors observed that compared to the wild-type plant (WT), the mutant increased the duration and reduced the amplitudes of electrical signals in sieve tube cells. Additionally, the expression of the JA pathway marker gene JAZ10 in leaves distal to wounds was enhanced (Kumari et al., 2019).

The Mechanosensitive Ion Channel-Like 10 (MSL10) is a type of ion channel that responds to physical forces and membrane deformations to facilitate ion flow, particularly anions with a conductance of about 100 pS (Lee and Seo, 2022). One study demonstrating its importance in plant electrical signaling under insect predation utilized *MSL10* mutants, which exhibited a significant reduction in the duration of SWP in systemic leaves. This reduction also led to decreased Ca^{2+} waves and expression of *JASMONATE-ZIM-DOMAIN PROTEIN 10* in distal tissues of *msl10* mutants (Moe-Lange et al., 2021).

Lee and Seo (2022) in their review work on the coordinating role of electrical signals in wound-induced systemic responses depicted a scheme representing local signaling and in distant tissues. In general, at the site of injury (Figure 1A), wounding triggers an elevation in apoplastic glutamate concentration, leading to Glutamate Receptor Like (GLR)-dependent Ca^{2+} influx and membrane potential depolarization. GLR proteins are presumed to function as tetramers. Subsequent to depolarization, membrane repolarization necessitates the activity of H^{+}-ATPase AHA1. The locally generated signals propagate to more distant parts through vascular tissues in association with a Ca^{2+} waves. Alterations in turgor pressure impact membrane tension, triggering the activation of the heptameric anion channel, MSL10, which induces membrane depolarization through anion efflux. This membrane depolarization fully activates GLRs in response to injury. GLR-dependent Ca^{2+} influx potentially activates MSL10 concurrently with the deactivation of AHA1. The synergistic propagation of Ca^{2+} waves and SWPs promotes JA biosynthesis to induce defense responses in systemic tissues (Figure 1B).

In addition to the generation and propagation scheme mentioned by Lee and Seo (2022), we will also detail other models and components that are being elucidated, aiming to complement the current understanding of electrical signaling during herbivory.

Overall, the models found in the literature share common components and mechanisms, but there are some differences depending on the authors. A more comprehensive theoretical model, developed by Kloth and Dicke (2022), based on current molecular knowledge, considers the involvement of other components, such as cyclic nucleotide-gated channels (CNGCs), auto-inhibited Ca^{2+}-ATPases (ACAs), and two-pore channel 1 (TPC1). The calcium-permeable channels TPC1, CNGC19, ACA10, and ACA12 are also implicated in controlling systemic calcium waves. ACA10 and ACA12 play a role in the recovery of excitability after prolonged herbivory. Proton pumps, such as AHA1, determine the duration of membrane potential, but can also contribute to eliciting depolarizations (Kloth and Dicke, 2022).

The cyclic nucleotide-gated channels (CNGCs) are recognized for their role in modulating calcium levels in response to various stressors (Meena et al., 2019). Among the CNGC family, CNGC19 stands out due to its significant and rapid upregulation in both local and systemic leaves, suggesting its specific involvement in herbivory perception responses (Breeze, 2019). The absence of CNGC19 function results in diminished defense mechanisms against herbivory. Studies involving the cngc19 mutant have revealed abnormal and reduced intravascular Ca^{2+} fluxes, which impact electrical signaling processes (Meena et al., 2019).

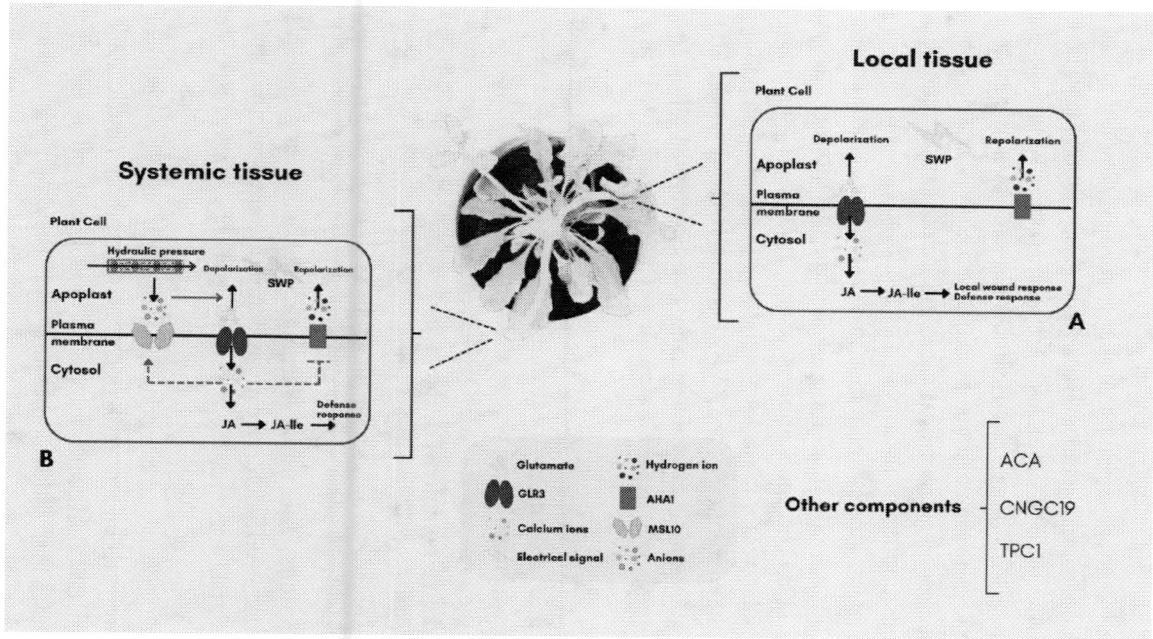

Figure 1. Scheme representing elements involved in local and systemic electrical signaling generation and propagation. Adaptation of the model and figure from Lee and Seo (2022). ACA: auto-inhibited Ca^{2+}-ATPases; CNGC19: cyclic nucleotide-gated channels 19; TPC1: two-pore channel 1. SWP: slow wave potential. JA: jasmonic acid; JA-lle: jasmonic acid-isoleucine (active form).

Another component involved in calcium dynamics during herbivory is the auto-inhibited Ca^{2+}-ATPases (ACAs), which are essential elements of the broader Ca^{2+} transport and signaling machinery. To investigate the influence of this component on electrical signaling during herbivory, Fotouhi et al., (2022) conducted a study using mutants in *ACA* genes. They found that when *A. Thaliana* P-Type II Ca^{2+}-ATPase mutants were mechanically wounded, all mutants tested exhibited leaf-to-leaf electrical signals. However, when the auto-inhibited Ca^{2+}-ATPase double mutant *aca10 aca12* was attacked by *Spodoptera littoralis* caterpillars, electrical signaling failed catastrophically, leading to the rapid consumption of these plants by the insects.

TPC1 ia a vacuolar ion chanel ubiquitous in the vacuolar membrane of terrestrial plants which transports small mono- and divalent cations (Pottosin and Dobrovinskaya, 2022). The application of pharmacological agents in *A. Thaliana* subjected to salt stress, such as La^{3+} and Ruthenium, inhibit calcium-release pathways and the disruption of the TWO PORE CHANNEL1 (TPC1) gene resulted in a substantial drop in wave speed, demonstrating the importante of this component in systemic siganling (Choi et al., 2014; Evans et al., 2016).

Choi et al. (2017) propose a model where a trio of messengers act collectively for systemic signaling. Local stresses, such as insect-induced injury, trigger changes in membrane potentials, increases in cytosolic Ca^{2+} concentration, and activation of RBOHD-mediated oxidative burst (respiratory burst oxidative homolog D), the latter leading to enhanced ROS accumulation. The interaction between the signals results in the propagation of a wave that rapidly spreads throughout the plant in a self-sustaining manner, traversing the extracellular apoplast and/or symplastically through plasmodesmata connections. This initial warning signal acts as a precursor, playing a necessary but insufficient role in systemic acquired acclimation. This initial warning signal activates the core environmental stress response genes.

Farmer et al. (2020) proposed a model for the propagation of electrical signals in Arabidopsis plants under herbivory, considering both physical and chemical activators of SWP. According to this model, caterpillar-induced wounding leads to a loss of phloem pressure and tension in the xylem. Consequently, pressure changes in the vessels facilitate the transport of elicitors into and along the xylem, where they interact with cells, promoting further depolarization. This initial stage is primarily non-electrical. Subsequently, the electrical phase of SWP propagation begins, marked by an increase in cytosolic calcium levels and the synthesis of jasmonate from precursors produced by LOX6. The resulting jasmonates are then exported to activate defense responses. The authors emphasize the importance of both xylem and phloem in SWP propagation. Recently, the same group proposed osmoelectric siphon mechanisms for elicitor transport, which are based on membrane depolarization inducing cell water shedding into the apoplast, followed by membrane repolarization and water uptake. For minor injuries, the authors suggested that broken cells release elicitors of membrane depolarization, which spread to neighboring cells. When these cells depolarize, they lose water, leading adjacent cell walls to absorb both water and elicitors, thereby enhancing their distribution in the apoplast. Eventually, the membranes repolarize, allowing cells to uptake water again. This process results in the radial spread of 'wet' apoplasm from small wounds. Once dispersed, the process ceases when the elicitor concentration decreases or decays, thereby triggering jasmonate synthesis (Gao and Farmer, 2023).

Ricca's factors, named after the researcher who first described them, are low molecular mass compounds believed to play a crucial role in initiating defense responses in plants. These compounds are thought to be released by damaged tissues, particularly in response to insect feeding or other forms of herbivory. Ricca's factors are hypothesized to act as elicitors, stimulating the plant's defense mechanisms, which include the production of various chemical compounds and the initiation of electrical signals such as SWPs. Although the precise chemical composition of Ricca's factors remains unclear, they are considered vital mediators of plant defense responses to herbivory (Gao et al., 2023).

To gain insight into Ricca's factors, Gao and colleagues (2023) conducted a study that uncovered a mechanism wherein organ-to-organ protein transport plays a significant role in electrical signaling. The study investigated mediators of leaf-to-leaf electrical signaling in Arabidopsis thaliana and identified them as b-THIOGLUCOSIDE GLUCOHYDROLASE 1 and 2 (TGG1 and TGG2). SWP propagation from insect feeding sites was markedly diminished in *tgg1 tgg2* mutants, accompanied by a reduction in wound-response cytosolic Ca^{2+} increases in these plants. Introducing recombinant TGG1 into the xylem induced membrane depolarization and Ca^{2+} transients similar to the wild type. Additionally, TGGs catalyze the deglucosidation of glucosinolates. Metabolite profiling revealed rapid wound-induced breakdown of aliphatic glucosinolates in primary veins. Through in vivo chemical trapping, evidence emerged for the involvement of short-lived aglycone intermediates, generated by glucosinolate hydrolysis, in SWP membrane depolarization.

There are several techniques employed to analyze the electrical signals generated and propagated in plants under insect herbivory. In the next section, we will discuss how we can acquire these electrical signals and some techniques for studying this data.

5. Data Acquisition and Analysis of Electrical Signals

Electricity is a phenomenon whose understanding has significantly transformed the contemporary world. The intricate and fascinating history of electricity, magnetism, and electromagnetism, fraught with gaps and controversies, dates back to the earliest days of scientific inquiry into the subject. Since the inception of this research, some scholars have contemplated the influence and role of electricity in living organisms. These conjectures were invigorated by Lemström (1904), who explored the stimulating effects of electrostatic fields on plant growth, both within and outside Faraday cages. Investigation into the potential applications of electricity persists to this day (McCaig et al., 2005; Goldsworthy, 2006).

However, the consolidation of electrophysiology as an effective scientific field occurred years after Luigi Galvani's discoveries in the 18th century, through the continuation of his studies by Carlo Matteucci and Emil du Bois-Reymond. These researchers observed a potential difference between injured and intact areas of a nerve, which they termed "injury potential." Furthermore, they found that stimulation of the nerve, whether mechanically or electrically, resulted in a disturbance in voltage, reducing and even reversing the voltage difference of the "injury potential." This phenomenon, now known as membrane depolarization, culminated in the inversion of the membrane potential, called the action potential (AP), responsible for muscular movement. These discoveries conferred greater relevance and robustness to electrophysiology (du Bois-Reymond, 1948; Stahlberg, 2006a,c).

Subsequently, with the demonstration that all cells, including those of plants, were enclosed by semipermeable membranes, and the consolidation of the cell theory, it became possible to mathematically model both resting potential and action potential based on the potassium gradient and its dissipation at the time of action potential generation. This modeling was developed by Wilhelm Nernst and Julius Bernstein in their works on the electromotive effectiveness of ions (1889) and the general properties of electrical excitation of nerves (1912), respectively.

The ability to rapidly determine plant stress, stimuli, and current physiological state through the classification of plant electrical signals is a tool akin to that used by physicians in diagnosing diseases in patients, as exemplified by electrocardiograms (Huang et al., 2010). In this context, various measurement techniques have been developed and refined to capture and analyze these signals accurately and efficiently.

One of the fundamental approaches in measuring electrical signals in plants is electrography. This technique involves placing electrodes directly on the surface of the plant to record variations in electrical potential over time. Electrography is particularly useful for monitoring changes induced by external stimuli such as light, temperature, and biotic or abiotic stresses (Saraiva et al., 2017). However, this technique presents some challenges such as limitations in spatial resolution, which can be restrictive, especially when investigating electrical events at the cellular level. Electrography may be sensitive to external interferences such as electromagnetic fields and environmental variations, which can impact the quality of the obtained data. Interpreting data generated by electrography can be complex and requires advanced analytical skills. The need for careful processing may pose a challenge for less experienced researchers.

Another relevant method is the intracellular microelectrode, which allows for the direct measurement of membrane potential in individual plant cells. This technique provides detailed insights into cellular electrical dynamics and has been widely used in studies on ion transport, osmotic regulation, and responses to specific stimuli. Intracellular microelectrodes offer high spatial resolution, enabling direct measurement of membrane potential. This allows for detailed analysis of electrical events at the cellular level. The technique enables precise assessment of electrical responses in specific cells, which is essential for understanding heterogeneity in cellular responses within a plant tissue or organ. Microelectrodes are effective in investigating ion transport processes, providing crucial insights into ion balance regulation and underlying mechanisms. However, the insertion of microelectrodes into plant cells is an invasive process that may alter the cell's normal conditions and influence physiological responses. This invasiveness can limit the application of the technique in long-term studies or in plants sensitive to disturbances. Precise manipulation of microelectrodes requires advanced technical skills. Improper insertion or mishandling can compromise the quality of obtained data. Given the intracellular nature of the technique, the number of cells that can be simultaneously monitored is limited. This may restrict the ability to study coordinated responses in large cell arrays (Walker and Graham, 1987).

In the context of plant electrophysiology, patch clamp techniques also stand out. Adapted from their application in animal cells, these techniques enable detailed assessment of the ion channels present in plant cell membranes, providing crucial insights into ion transport mechanisms and their regulation. Patch clamp technique offers high temporal and spatial resolution, allowing for precise measurement of electrical events in individual plant cells. This enables detailed study of ionic currents and membrane potentials. Patch clamp is capable of evaluating ionic currents associated with specific channels present in the cell membrane. This is crucial for understanding ion transport mechanisms and channel regulation. The technique allows assessment of electrical responses to specific stimuli, contributing to understanding how plants perceive and respond to changes in their environment. Similarly to animal cells, patch clamp in plants can be employed to study specific potentials, providing valuable information on intercellular communication. However, patch clamp technique is technically challenging and requires advanced skills for electrode manipulation. The learning curve is steep, which may limit its application to inexperienced researchers. Patch clamp application involves perforation of the cell membrane, which is intrusive and may result in cell damage. This invasiveness can influence physiological responses and limit the feasibility of long-term experiments. The technique is generally applied to individual cells, resulting in a low sampling rate when compared to methods monitoring larger cell populations. Patch clamp is more challenging to apply to complex plant tissues due to cellular heterogeneity. This may limit the ability to study coordinated responses at broader levels. The equipment required for patch clamp technique is specialized and can be costly, posing a challenge for laboratories with limited resources (Demidchik et al., 2006; Wegner et al., 2011; Gradogna and Carpaneto, 2022).

Aphid electrodes represent yet another intracellular technique facilitating the direct measurement of electrical signals in plants during interactions with insects. These electrodes are implanted into aphids' bodies as they feed on plants, capturing the electrical signals produced in plant cells in response to insect activity. This method offers a more comprehensive understanding of plant defense mechanisms against pests and provides valuable insights into insect adaptations aimed at circumventing these defenses (Hedrich et al., 2016). This technique has been integrated with electropenetography and finds extensive use in entomological studies, particularly for investigating insect feeding behavior (Lucini and Panizzi, 2020). In this system, the sucking insect becomes part of a simple electrical circuit along with its host plant. The circuit completes when the insect inserts its piercing-sucking mouthparts (stylets) into the electrified plant tissue. The device records stylet activities, such as ingestion, as waveforms; analyzing each waveform allows for the quantification of different stylet activities (Lucini and Panizzi, 2020; Walker, 2000).

We can mention other less commonly used techniques such as specific ion sensors, which have significantly contributed to the understanding of real-time ionic fluxes. Another research avenue includes electrochemical spectroscopy, which excels in assessing metabolic and redox processes in plants. In summary, the diversity of techniques for measuring electrical signals in plants reflects the complexity and relevance of this research field. The strategic combination of these approaches allows for a deeper understanding of the electrical mechanisms underlying physiological processes, contributing to significant advances in agriculture, biotechnology, and fundamental understanding of plant life.

However, the acquisition of the signal represents only the initial stage of the analytical process. The intrinsic biological nature of this signal imposes considerable complexity on its analysis. Over the years, significant advances have been made in understanding this phenomenon, with notable progress achieved through the conception of the term "electrome." Arnold De Loof (De Loof, 2016) developed and introduced this concept to describe the summation of all self-generated electrical activity by a biological compartment over a certain time interval. This definition establishes an analogy with collective characteristics observed at other biological levels, such as the genome, proteome, and metabolome. According to De Loof, the electrome represents the fundamental electrical dimension of life, without which life itself would be unviable (De Loof, 2016). With this understanding, the analysis of electrophysiological signals has evolved into a collective perspective, directing focus not only to individual cellular signals. However, in approaching the signal in this manner, a considerable challenge arises regarding multi-body interaction. The "Three-Body Problem," a mathematical and physical challenge, involves predicting the motion of three celestial bodies interacting gravitationally with each other, without general analytical solutions. The inherent complexity of this problem arises from its extreme sensitivity to initial conditions, resulting in chaotic and unpredictable behaviors over time. In summary, multi-body interaction confers upon the electrome an emergent and chaotic character in the electrical dynamics of plant cells (Souza, 2004; Souza, 2005; Souza, 2015).

The concept of chaos, introduced in the context of dynamical systems, denotes a state of complex apparent disorder and unpredictability, characterized by nonlinear and often seemingly random behaviors. This approach challenges the traditional perspective of determinism and predictability in dynamical systems, highlighting the possibility of chaotic behaviors emerging from fundamental deterministic laws. Understanding chaos proves to be of crucial importance in exploring complex phenomena in natural systems, playing a significant role in advancing the theory of dynamical systems and practical applications across various scientific disciplines.

Given this chaotic nature, electrome analysis cannot be adequately summarized by traditional approaches such as simple statistics, averages, or frequentist comparisons. In this context, Machine Learning (ML) emerges as a promising alternative for testing hypotheses, such as the suggestion that fruit, despite being a significant sink, is capable of informing the mother plant about a potential attack (Reissig et al., 2021). However, ML presents a notable limitation, the so-called "black box," which, despite its ability to classify physiological states with great accuracy, does not offer a clear understanding of the underlying decision-making process.

To overcome this challenge, the Time Dispersion Analysis of Features (TDAF) method emerges. In electrome analysis, we identify a fundamental issue: the temporal dimension. Unlike chemical analyses aimed at understanding the plants physiology, electrome study incorporates not only the spatial dimension but also the temporal dimension (Saraiva et al., 2017; Souza et al., 2017). This implies that time must be considered integrally in this analysis, as biological entities are constantly undergoing changes. Unlike static data, there is constant dynamics in their fundamental characteristics, and any average or statistic will never fully reflect the complexity of the system.

Based on this premise, TDAF is grounded in dividing the temporal series into the smallest possible unit, considering the limitations of the measurement equipment. Utilizing dispersion theory (Jorgensen, 1997), TDAF analyzes the dataset as a whole, on a minute-by-minute basis. The distinctiveness of this analysis lies in the fact that, by utilizing previously obtained information, it's possible to associate each dispersion analysis with its respective moment in time. This procedure results in the creation of a new temporal series (Costa et al., 2023), providing a more robust analytical approach for understanding the electrome in its temporal and spatial complexity. Next, we will discuss how the application of data acquisition and analysis techniques can be used in practice, as well as prospects for studying electrical signals in plants under insect predation.

Conclusion and Future Perspectives

Plant electrophysiology is increasingly gaining recognition as a tool that can aid not only in monitoring abiotic stresses in plants but also in the pre-diagnosis of biotically induced diseases (Simmi et al., 2023) and the detection of insect predation (Reissig et al., 2021). As data analysis techniques continue to improve, it becomes possible to gain more insights into the information contained within the electrical signals. It may soon be possible, for instance, to identify the species of fungus or insect that is harming the plant. This would enable more appropriate and precise treatment, reducing production losses and minimizing the use of agricultural inputs.

In their research, Pachú et al., (2021) observed that both Bt (genetically modified pest-resistant) and non-Bt cotton plants emit variation potential/slow wave potential electrical signals when attacked by aphids. These signals reflect distinct responses in terms of plant perception and aphid behavior, with evidence of delayed signal amplitude and significantly elevated levels of Cry1F protein observed on Bt cotton plants under higher insect densities. They indicate that their results may serve as a roadmap for future studies aimed at elucidating the factors involved in stress resistance and plant defense processes, thereby aiding in the development of effective strategies in integrated pest management.

In their study on electrical signals generated in fruits under herbivory and their transmission throughout the plant, Reissig and colleagues (2021) observed that following an attack by the *Helicoverpa armigera* caterpillar, it was possible to classify the generated signal using Machine Learning techniques. They suggested that this knowledge could be leveraged in the future to identify and differentiate the type of infestation affecting a crop. Another significant contribution of this work was the demonstration that not only predation on leaves but also attacks on fruits elicit a response in the entire plant, as detected by electrophysiological techniques.

In addition to a horizon of benefits that the utilization of this knowledge can bring for more sustainable food production, research into how insect predation on plants affects plant electrical activity expands fundamental scientific understanding, facilitating the comprehension of the mechanisms underlying local and systemic signaling, as well as plant cognitive processes, which enable plants to survive in constantly changing environments.

Acknowledgments

The authors are grateful to LACEV (Laboratory of Plant Cognition and Electrophysiology, UFPel, Brazil) for the support provided. This study was supported in part by the Coordenação de Aperfeiçoamento de Pessoal de Nível Superior, Brasil (CAPES, Brazil), Finance Code 001. DEN acknowledge the National Council for Scientifc and Technological Development (CNPq, Brazil) for the Productivity Research Grant provided.

Disclaimer

The authors declare that they do not have any known competing financial interests or personal relationships that could have influenced the work reported in this paper.

References

Aartsma Y., Bianchi F. J. J., A. van der Werf W., Poelman E. H., Dicke M. Herbivore-induced plant volatiles and tritrophic interactions across spatial scales. *New Phytologist* (2017) 216:1054–1063.

Acevedo F. E., Rivera-Vega L. J., Chung S. H., Ray S., Felton G. W. Cues from chewing insects - the intersection of DAMPs, HAMPs, MAMPs and effectors. *Current Opinion in Plant Biology* (2015) 26:80-6.

Ashour M., Wink M., Gershenzon J. Biochemistry of terpenoids: monoterpenes, sesquiterpenes and diterpenes. *Biochemistry of Plant Secondary Metabolism Annual Plant Reviews* (2010) 40: 258–303.

Bakker P. A. H. M. Induced systemic resistance by beneficial microbes. *Annual Review of Phytopathology* (2014) 52: 347–375.

Baluska F., Mancuso S. Deep evolutionary origins of neurobiology: turning the essence of 'neural' upside-down. Co*gnitive Processing* (2009b) 2(1):60–65.

Baluska F., Mancuso S. Individuality, self and sociality of vascular plants. *Philosophical Transactions of the Royal Society of London* (2021) 376:20190760.

Baluska F., Mancuso S. Ion channels in plants: from bioelectricity, via signaling, to behavioral actions. *Plant Signaling & Behavior* (2013a) 8(1):e23009

Baluska F., Mancuso S. Root apex transition zone as a oscillatory zone. *Frontiers in Plant Science* (2013b) 4(354):1–15

Baluska F., Mancuso S., Volkamann D. *Communication in plants: neuronal aspects of plant life*. Springer, Berlin (2006).

Baluska F., Mancuso, S. Plant neurobiology: from sensory biology, via plant communication, to social plant behavior. *Cognitive Processing* (2009a) 10: 3-7.

Bonaventure G. Perception of insect feeding by plants. *Plant Biology* (2012) 14:872–880.

Breeze E. State of (In)flux: Action of a CNGC Ca^{2+} Channel in Defense against Herbivory. *Plant Cell* (2019) 31(7):1423-1424.

Brenner E. D., Stahlberg R., Mancuso S., Vivanco J. M., Baluska F., van Volkenburgh E. Plant neurobiology: an integrated view of plant signaling. *Trends in Plant Science* (2007) 11(8):413–419.

Canales J., Henriquez-Valencia C., Brauchi S. The integration of electrical signals originating in the root of vascular plants. *Frontiers in Plant Science* (2018) 8:2173.

Choi W. G., Miller G., Wallace I., Harper J., Mittler R., Gilroy S. Orchestrating rapid long-distance signaling in plants with Ca^{2+}, ROS and electrical signals. *The Plant Journal* (2017) 90(4):698-707.

Choi W. G., Toyota M., Kim S. H., Gilroy S. Salt stress-induced Ca^{2+} waves are associated with rapid, long-distance root-to-shoot signaling in plants. *Proceedings of the National Academy of Sciences* (2014) 111(17):6497-6502.

Costa Á. V. L., Oliveira T. F. d. C., Posso D. A., Reissig G. N., Parise A. G., Barros W. S., Souza G. M. Systemic signals induced by single and combined abiotic stimuli in Common Bean Plants. *Plants* (2023) 12(4):924.

Cruaud A., Rønsted N., Chantarasuwan B., Chou L. S., Clement W. L., Couloux A., Cousins B., Genson G., Harrison R. D., Hanson P. E., Hossaert-McKey M., Jabbour-Zahab R., Jousselin E., Kerdelhué C., Kjellberg F., Lopez-Vaamonde C., Peebles J., Peng Y. Q., Pereira R. A., Schramm T., Ubaidillah R., van Noort S., Weiblen G. D., Yang D. R., Yodpinyanee A., Libeskind-Hadas R., Cook J. M., Rasplus J. Y., Savolainen V. An extreme case of plant-insect codiversification: Figs and fig-pollinating wasps. *Systematic Biology* (2012) 61:1029-1047.

Culliney T. W. Crop losses to arthropods. In Integrated Pest Management: Pesticide Problems, Vol. 3 (D. Pimentel, and R. Peshin, eds), pp. 201–225. Dordrecht, Netherlands: Springer, 2014.

De Bruxelles G. L., Roberts, M. R. Signals regulating multiple responses to wounding and herbivores. *Critical Reviews in Plant Sciences* (2001) 20:487–521.

De Loof A. The cell's self-generated "electrome": the biophysical essence of the immaterial dimension of Life? *Communicative and Integrative Biology* (2016) 9:e1197446.

De Vos M., Jander G. Volatile communication in plant-aphid interactions. *Current Opinion in Plant Biology* (2010) 13:366–371.

Dean R. A., Kuć J. Induced systemic protection in plants. *Trends in Biotechnology* (1985) 3:125–129.

Dejean A., Petitclerc F., Roux O., Orivel J., Leroy C. Does exogenic food benefit both partners in an ant-plant mutualism? The case of *Cecropia obtusa* and its guest *Azteca* plant-ants. *Comptes Rendus Biologies* (2012) 335:214-219.

Demidchik V., Sokolik A., Yurin V. Electrophysiological characterization of plant cation channels. In: *Plant electrophysiology: theory and methods.* Berlin, Heidelberg: Springer Berlin Heidelberg. p. 173-185, 2006.

Dicke M., van Poecke R. M. P., de Boer J. G. Inducible indirect defence of plants: from mechanisms to ecological functions. *Basic and Applied Ecology* (2003) 4:27–42.

Du Bois-Reymond. *Untersuchungen über thierische Elektrizität*, Reimer, Berlin, 1948.

Erb M., Meldau S., Howe G. A. Role of phytohormones in insect-specific plant reactions. *Trends in Plant Science* (2012) 17:250–259.

Evans M. J., Choi W. G., Gilroy S., Morris R. J. A ROS-Assisted Calcium Wave Dependent on the AtRBOHD NADPH Oxidase and TPC1 Cation Channel Propagates the Systemic Response to Salt Stress. *Plant Physiol*ogy (2016) 171(3):1771-84.

Farmer E. E., Gao Y. Q., Lenzoni G., Wolfender J. L., Wu Q. Wound- and mechanostimulated electrical signals control hormone responses. *New Phytologist* (2020) 227:1037-1050.

Fotouhi N., Fischer-Stettler M., Lenzoni G., Stolz S., Glauser G., Zeeman S. C., Farmer E. E. ACA pumps maintain leaf excitability during herbivore onslaught. *Current Biology* (2022) 6;32(11):2517-2528.e6.

Fromm J., Lautner S. Electrical signals and their physiological significance in plants. *Plant, Cell & Environment* (2007) 30:249–257.

Gao Y. Q., Farmer E. E. Osmoelectric siphon models for signal and water dispersal in wounded plants. *Journal of Experimental Botany* (2023) 74(4):1207–1220.

Gao Y. Q., Jimenez-Sandoval P., Tiwari S., Stolz S., Wang J., Glauser G., Santiago J., Farmer E. E. Ricca's factors as mobile proteinaceous effectors of electrical signaling. *Cell* (2023) 30;186(7):1337-1351.e20.

Gaskett A. C. Orchid pollination by sexual deception: pollinator perspectives. *Biological reviews of the Cambridge Philosophical Society* (2011) 86(1):33-75.

Goldsworthy A. Effects of Electrical and Electromagnetic Fields on Plants and Related Topics, p 247-268. In Volkov, A. *Plant Electrophysiology: Theory and methods.* Springer-Verlag, Berlin, Germany, 512 p, 2006.

Gradogna A., Carpaneto A. Electrophysiology and fluorescence to investigate cation channels and transporters in isolated plant vacuoles. *Stress Biology* (2022) 2(1):42.

Grunseich, J. M., Morgan N. Thompson, Allison A. Hay, Zachary Gorman, Michael V. Kolomiets, Micky D. Eubanks, Anjel M. Helms Risky roots and careful herbivores: Sustained herbivory by a root-feeding herbivore attenuates indirect plant defences. Functional Ecology (2020) 34(9):1779–1789.

Haas J., Lozano E. R., Haida K. S., Mazaro S. M., de Souza Vismara E., Poppy G. M. Getting ready for battle: do cabbage seeds treated with jasmonic acid and chitosan affect chewing and sap-feeding insects? *Entomologia Experimentalis et Applicata* (2018) 166:412–419.

Hedrich R., Salvador-Recatalà, V., Dreyer, I. Electrical wiring and long-distance plant communication. *Trends in Plant Science* (2016) 21.5: 376-387.

Heil M., Bueno J. C. S. Within-plant signaling by volatiles leads to induction and priming of an indirect plant defense in nature. *Proceedings of the National Academy of Sciences* (2007) 104:5467–5472.

Howe A. G., Jander G. Plant immunity to insect herbivores. *Annual Review of Plant Biology* (2008) 59:41–66.

Hu C., Wu S., Li J., Dong H., Zhu C., Sun T., Hu Z., Foyer C. H., Yu J. Herbivore-induced Ca^{2+} signals trigger a jasmonate burst by activating ERF16-mediated expression in tomato. *New Phytologist* (2022) 236(5): 1796-1808.

Huang L., Wang Z. Y., Zhao L. L., Zhao D. J., Wang C., Xu Z. L., Hou R. F., Qiao X. J. Electrical signal measurement in plants using blind source separation with independent component analysis. *Computers and Electronics in Agriculture* (2010) 71:S54–S59.

Huber A. E., Bauerle T. L. Long-distance plant signaling pathways in response to multiple stressors: the gap in knowledge. *Journal of Experimental Botany* (2016) 67:2063–2079.

Jorgensen B. *The Theory of Dispersion Models* (Chapman & Hall/CRC Monographs on Statistics & Applied Probability). 1. ed. London: Chapman and Hall/CRC, 1997.

Karabourniotis G., Liakopoulos G., Nikolopoulos D., Bresta P. Protective and defensive roles of non-glandular trichomes against multiple stresses: structure-function coordination. *Journal of Forestry Research* (2020) 31:1–12.

Kloepper J. W., Tuzun S., Kuć J. A. Proposed definitions related to induced disease resistance. *Biocontrol Science and Technology* (1992) 2:349–351.

Kloth K. J., Dicke M. Rapid systemic responses to herbivory. *Current Opinion in Plant Biology* (2022) 68:102242.

Kortbeek R. W. J., Galland M. D., Muras A., van der Kloet F. M., André B., Heilijgers M., van Hijum S. A. F. T., Haring M. A., Schuurink R. C., Bleeker P. M. Natural variation in wild tomato trichomes; selecting metabolites that contribute to insect resistance using a random forest approach. *BMC Plant Biology* (2021) 21:315.

Kuć J. Concepts and direction of induced systemic resistance in plants and its application. *European Journal of Plant Pathology* (2001) 107:7–12.

Kumari A., Ch etelat A., Nguyen C. T., Farmer E. E. Arabidopsis H+-ATPase AHA1 controls slow wave potential duration and wound-response jasmonate pathway activation. *Proceedings of the National Academy of Sciences* (2019) 116:20226–20321.

Labandeira C. C. The four phases of plant-arthropod associations in deep time. *Geologica Acta* (2006) 4:409-438.

Lazebnik J., Frago E., Dicke M., van Loon J. J. Phytohormone mediation of interactions between herbivores and plant pathogens. *Journal of Chemical Ecology* (2014) 40:730–741.

Lee B., Lee S., Ryu C. M. Foliar aphid feeding recruits rhizosphere bacteria and primes plant immunity against pathogenic and non-pathogenic bacteria in pepper. *Annals of Botany* (2012) 110:281–290.

Lee K., Seo P. J. Wound-induced systemic responses and their coordination by electrical signals. *Frontiers in Plant Science* (2022) 18(13):880680.

Lemström S. *Electricity in agriculture and horticulture*. The Electrician Printing & Publishing, London. 1904

Liu Q., Li S., Ding W. Aphid-induced tobacco resistance against *Ralstonia solanacearum* is associated with changes in the salicylic acid level and rhizospheric microbial community. *European Journal of Plant Pathology* (2020) 157:465–483.

Lucini T., Panizzi A. R. Electropenetrographic comparison of feeding behavior of *Dichelops furcatus* (Hemiptera: Heteroptera: Pentatomidae) on soybean and spring cereals. *Journal of Economic Entomology* (2020) 13;113(4):1796-1803.

Maffei M. E., Mithofer A., Boland W. Before gene expression: early events in plant-insect interaction. *Trends Plant in Science.* (2007) 12:310–316.

Maffei M., Bossi S., Spiteller D., Mithofer A., Boland W. Effects of feeding *Spodoptera littoralis* on lima bean leaves. I. Membrane potentials, intracellular calcium variations, oral secretions, and regurgitate components. *Plant of Physiology* (2004) 134:1752–1762.

Marhavý P., Andrzej Kurenda, Shahid Siddique, et al., Single-cell damage elicits regional, nematode-restricting ethylene responses in roots. *The EMBO journal* (2019) 38(10):e100972.

Martinez de Ilarduya O., Xie Q., Kaloshian I. Aphid-induced defense responses in Mi-1-mediated compatible and incompatible tomato interactions. *Molecular Plant-Microbe Interactions Journal* (2003) 16:699–708.

McCaig C. D., Rajnicek A. M., Song B., Zhao M. Controlling cell behavior electrically: Current views and future potential. *Physiological Reviews* (2005) 85:943–978.

Meena M. K., Prajapati R., Krishna D., Divakaran K., Pandey Y., Reichelt M., Mathew M. K., Boland W., Mithöfer A., Vadassery J. The Ca^{2+} Channel CNGC19 Regulates *Arabidopsis* defense against *Spodoptera h*erbivory. *Plant Cell* (2019) 31(7):1539-1562.

Mithöfer A., Boland W. Plant defence against herbivores: Chemical aspects. *Annual Review of Plant Biology* (2012) 63:431–450.

Moe-Lange J., Gappel N. M., Machado M., Wudick M. M., Sies C. S. A., Schott-Verdugo S. N., Bonus M., Mishra S., Hartwig T., Bezrutczyk M., Basu D., Farmer E. E., Gohlke H., Malkovskiy A., Haswell E. S., Lercher M. J., Ehrhardt D. W., Frommer W. B., Kleist T. J. Interdependence of a mechanosensitive anion channel and glutamate receptors in distal wound signaling. *Science Advances* (2021) 10;7(37):eabg4298.

Moran P. J., Thompson G. Molecular responses to aphid feeding in Arabidopsis in relation to plant defense pathways. *Plant Physiology* (2001) 125:1074–1085.

Mousavi S. A. R., Chauvin A., Pascaud F., Kellenberger S., Farmer E. E. Glutamate Receptor-Like Genes mediate leaf-to-leaf wound signalling. *Nature* (2013) 500:422–426.

Najdenovska E., Dutoit F., Tran D., Plummer C., Wallbridge N., Camps C., Raileanu L. E. Classification of plant electrophysiology signals for detection of Spider Mites infestation in tomatoes. *Applied Sciences* (2021) 11:1414.

Núñez-Farfán J., Fornoni J., Valverde P. L. The evolution of resistance and tolerance to herbivores. *Annual Review of Ecology, Evolution, and Systematics* (2007) 38:541–566.

Pachú J. K. S., Macedo F. C. O., Malaquias J. B., Ramalho F. S., Oliveira R. F., Franco F. P., Godoy W. A. C. Electrical signalling on Bt and non-Bt cotton plants under stress by *Aphis gossypii. PLoS One* (2021) 8;16(4):e0249699.

Pieterse C. M. J., Zamioudis C., Berendsen R. L., Weller D. M., Van Wees S. C. M., Ollerton J., Winfree R., Tarrant S. How many flowering plants are pollinated by animals? *Oikos* (2011) 120(3):321–326.

Pottosin I., Dobrovinskaya O. Major vacuolar TPC1 channel in stress signaling: what matters, K^+, Ca^{2+} conductance or an ion-flux independent mechanism? *Stress Biology* (2022) 2:31.

Qi X., Chen M., Liang D., Xu Q., Zhou F., Chen X. Jasmonic acid, ethylene and ROS are involved in the response of cucumber (*Cucumis sativus* L.) to aphid infestation. *Scientia Horticulturae* (2020) 269:109421.

Reissig G. N., Oliveira T. F. C., Oliveira R. P., Posso D. A., Parise A. G., Nava D. E., Souza G. M. Fruit herbivory alters plant electrome: Evidence for fruit-shoot long-distance electrical signaling in tomato plants. *Frontiers in Sustainable Food Systems* (2021) 5:657401.

Rubil N., Kalachova T., Hauser T. P., Burketová L. Specialist aphid feeding causes local activation of salicylic and jasmonic acid signaling in Arabidopsis veins. *Molecular Plant-Microbe Interactions* (2022) 35:119–124.

Saijo Y., Loo E. P., Yasuda S. Pattern recognition receptors and signaling in plant–microbe interactions. *The Plant Journal* (2018) 93(4):592–613.

Saraiva G. F. R., Ferreira A. S., Souza G. M. Osmotic stress decreases complexity underlying the electrophysiological dynamic in soybean. *Plant Biology* (2017) 19(5):702-708.

Schäfer M., Fischer C., Meldau S., Seebald E., Oelmüller R., Baldwin I. T. Lipase activity in insect oral secretions mediates defense responses in Arabidopsis. *Plant Physiology* (2011) 156:1520–34.

Schatz B., Nicolas Sauvion, Finn Kjellberg. Plant–Insect Interactions: A Palaeontological and an Evolutionary Perspective. In: *Advances in Botanical Research.* [s.l.] Elsevier (2017) 81:1–24.

Schultz J. C. Plant responses induced by herbivores. *Trends in Ecology and Evolution* (1988), 3:45–49.

Shepherd V. A. At the roots of plant neurobiology. In: Volkov A. G. (ed) *Plant electrophysiology*. Springer, Berlin, pp 3–43, 2012.

Shepherd V. A. From semi-conductors to the rhythms of sensitive plants: the research of J. C. Bose. *Cellular and Molecular Biology* (2005) 51:607–619.

Simmi F. Z., Dallagnol L. J., Almeira R. O., Dorneles K. R., Souza G. M. Barley systemic bioelectrical changes detect pathogenic infection days before the first disease symptoms. *Computers and Electronics in Agriculture* (2023) 209:107832.

Smith C. M., Clement S. L. Molecular bases of plant resistance to arthropods. *Annual Review of Entomology*. (2012) 57:309–328.

Souza G. M., Ferreira A. S., Saraiva G. F. R., Toledo G. R. A. Plant "electrome" can be pushed toward a self-organized critical state by external cues: Evidences from a study with soybean seedlings subject to different environmental conditions. *Plant Signaling & Behavior* (2017) 12(3):e1290040.

Souza G. M., Lüttge U. Stability as a phenomenon emergent from plasticity–complexity–diversity in eco-physiology. *Progress in Botany* (2015) 76:211–239.

Souza G. M., Oliveira R. F., Cardoso V. J. M. Temporal dynamics of stomatal conductance of plants under water deficit: can homeostasis be improved by more complex dynamics. *Arquivos de Biologia e Tecnologia* (2004) 47:423–431.

Souza G. M., Pincus S. M., Monteiro J. A. F. The complexity–stability hypothesis in plant gas exchange under water deficit. *Brazilian Journal of Plant Physiology* (2005) 17:363– 373

Stahl E., Hilfiker O., Reymond P. Plant-arthropod interactions: Who is the winner? *Plant Journal* (2018) 93:703–28

Stahlberg R. Historical Introduction to Plant Electrophysiology. In Volkov, A. *Plant Electrophysiology: Theory and methods*. Springer-Verlag, Berlin, Germany, (2006a) 512:3 – 14.

Stahlberg R. Historical Overview on Plant Neurobiology. *Plant Signaling & Behavior* (2006c) 1(1):6-8.

Stahlberg R., Cleland R. E., Van Volkenburgh, E. Slow wave potentials - a propagating electrical signal unique to higher plants. *Communication in plants: Neuronal Aspects of Plant Life* (2006b): 291-308.

Stavrinidou E., Gabrielsson R., Gomez E., Crispin X., Nilsson O., Simon D. T., Berggren M. Electronic plants. *Science Advances* (2015) 1(10):e1501136.

Sukhova E., Akinchits E., Sukhov V. Mathematical models of electrical activity in plants. *The Journal of Membrane Biology* (2017) 250:407–423.

Szechynska-Hebda M., Lewandowska M., Karpinski S. Electrical signaling, photosynthesis and systemic acquired acclimation. *Frontiers in Physiology* (2017) 8:684.

Thaler J. S., Humphrey P. T., Whiteman N. K. Evolution of jasmonate and salicylate signal crosstalk. *Trends in Plant Science* (2012) 17:260–270.

Thompson G. A., Goggin F. L. Transcriptomics and functional genomics of plant defence induction by phloem-feeding insects. *Journal of Experimental Botany* (2006), 57:755–766.

Toledo G. R. A., Parise A. G., Simmi F. Z., Costa A. V. L., Senko L. G. S., Debono M. W., et al., Plant electrome: the electrical dimension of plant life. *Theoretical and Experimental Plant Physiology* (2019) 31:21–46.

Toyota M., Spencer D., Sawai-Toyota S., Jiaqi W., Zhang T., Koo A. J., Howe G. A., Gilroy S. Glutamate triggers long-distance, calcium-based plant defense signaling. *Science* (2018) 14;361(6407):1112-1115.

Trebacz K., Dziubinska H., Krol E. Electrical signals in long-distance communication in plants. In: Balusˇka F, Mancuso S, Volkmann D (eds) *Communication in plants. Neuronal aspects of plant life* (2006) Springer, Berlin, pp 277–290.

Trewavas A. Intelligence, cognition, and language of green plants. *Frontiers in Psychology* (2016) 7:588.

Trewavas A. The foundations of plant intelligence. *Interface Focus* (2017) 7:20160098.

Van Loon L. C. Systemic induced resistance. In: Slusarenko, A. J., Fraser, R. S. S., van Loon, L. (Eds.), *Mechanisms of Resistance to Plant Diseases*. Kluwer Academic Publishers, Dordrecht, 521–574, 2000.

Vodeneev V. A., Katicheva L. A., Sukhov V. S. Electrical signals in higher plants: mechanisms of generation and propagation. *Biophysics* (2016a) 61(3):505–512.

Vodeneev V., Akinchits E., Sukhov V. S. Variation potential in higher plants: mechanisms of generation and propagation. *Plant Signaling & Behavior* (2016b) 10:e1057365.

Volkov A. G. *Plant electrophysiology: theory and methods*, 1st edn. Springer, Berlin, 2006.

Volkov A. G., Carrell H., Markin V. S. Biologically closed electrical circuits in venus flytrap. *Plant Physiology* (2009) 149:1661–1667.

Volkov A. G., Collins D. J., Mwesigwa J. Plant electrophysiology: pentachlorophenol induces fast action potentials in soybean. *Plant Science* (2000) 153:185–190.

Walker C. D., Graham R. D. A micro-electrode fast data capture technique: its use to examine plant cell electrophysiology in the wheat root. *Journal of Experimental Botany* (1987) 38(2):355-366.

Walker G. P. A beginner's guide to electronic monitoring of homopteran probing behavior, pp. 14–40. In G. P. Walker and E. A. Backus (eds.), *Principles and applications of electronic monitoring and other techniques in the study of homopteran feeding behavior*. Thomas Say Publications in Entomology, Entomological Society of America, Lanham, MD. 2000.

Walling L. L. The myriad plant responses to herbivores. *Journal of Plant Growth Regulation* (2000), 19:195–216.

Wegner L. H., Bianca Flickinger, Christian Eing, Thomas Berghöfer, Petra Hohenberger, Wolfgang Frey, Peter Nick. A patch clamp study on the electro-permeabilization of higher plant cells: Supra-physiological voltages induce a high-conductance, K^+ selective state of the plasma membrane. *Biochimica et Biophysica Acta (BBA)-Biomembranes* (2011) 1808(6):1728-1736.

Yan C., Fan M., Yang M., Zhao J., Zhang W., et al., Injury activates Ca^{2+} /calmodulin-dependent phosphorylation of JAV1-JAZ8-WRKY51 complex for jasmonate biosynthesis. *Molecular Cell* (2018) 70:136–37.

Yan C., Gao Q., Yang M., Shao Q., Xu X., Zhang Y., Luan S. Ca^{2+}/calmodulin-mediated desensitization of glutamate receptors shapes plant systemic wound signalling and anti-herbivore defence. *Nature Plants* (2024) 10(1):145-160.

Yan X., Wang Z., Huang L., Wang C., Hou R., Xu Z., Qiao X. Research progress on electrical signals in higher plants. *Progress in Natural Science* (2009) 19:531–541.

Zebelo S. A., Maffei M. E. Role of early signalling events in plant-insect interactions. *Journal of Experimental Botany* (2015) 66:435–448.

Zhou S., Jander G. Molecular ecology of plant volatiles in interactions with insect herbivores. *Journal of Experimental Botany* (2022) 73(2):449–462.

Zhu L., Guo J., Ma Z., Wang J., Zhou C. Arabidopsis transcription factor MYB102 increases plant susceptibility to aphids by substantial activation of ethylene biosynthesis. *Biomolecules* (2018) 8(2):39.

Zimmermann M. R., Maischak H., Mithoefer A., Boland W., Felle H. H. System potentials, a novel electrical long-distance apoplastic signal in plants, induced by wounding. *Plant Physiology* (2009) 149:1593–1600.

Zimmermann M., Mithofer A., Will T. et al., Herbivore-Triggered electrophysiological reactions: candidates for systemic signals in higher plants and the challenge of their identification. *Plant Physiology* (2016) 170:2407–2419.

Züst T., Agrawal A. A. Mechanisms and evolution of plant resistance to aphids. *Nature Plants* (2016) 2:15206.

Biographical Sketches

Gabriela Niemeyer Reissig

Affiliation:
Federal University of Pelotas (UFPel). Laboratory of Cognition and Plant Electrophysiology (LACEV).

Education:
Bachelor's degree in Food Chemistry (UFPel). Master's and PhD in Food Science and Technology (UFPel).

Business Address:
Federal University of Pelotas, Department of Botany. University Campus, Andre Dreyfuss Alley building 21. Jardim América. ZIP code: 96160000 - Capão do Leão, RS – Brazil.

Research and Professional Experience:
Postdoctoral researcher investigating fruit electrophysiology and cognitive processes in plants.

Publications from the Last 3 Years:
2023:

Costa, Á. V. L., Oliveira, T. F. C., Posso, D. A., Reissig, G. N., Parise, A. G., Barros, W. S., Souza, G. M. Systemic Signals Induced by Single and Combined Abiotic Stimuli in Common Bean Plants. *Plants,* v. 12, p. 924, 2023.

Reissig, G. N., Oliveira, T. F. C., Parise, A. G., Costa, Á. V. L., Posso, D. A., Rombaldi, C. V., Souza, G. M. Approximate entropy: a promising tool to understand the hidden electrical activity of fruit. *Communicative & Integrative Biology*, v. 16, p. 2195236, 2023.

Reissig, G. N., Basso, L. F., Ramos, L. P., de Souza, J. G. M., de Almeida, L. B. Inteligência E Eletrofisiologia Vegetal. *Expressa Extensão* (UFPEL), v. 28, p. 61-66, 2023.

Posso, D. A., Da-Silva, C. J., Shimoia, E. P., Martins, T. da S., Reissig, G. N., De Oliveira, A. C. B., Borella, J., Van Dongen, J. T., do Amarante, L. Root-hypoxia tolerance in soybean sister-lines plants indicates a better balance in energy use/dissipation and oxidative stress control. *Plant Stress*, v. 10, p. 100225, 2023.

Reissig, G. N., Oliveira, T. F. C., Bragança, G. C. M., Rodrigues, R. da S., Barbosa, F. da F. Fermented vegetables and fruits as vitamin B12 sources: An overview. *International Food Research Journal,* v. 30, p. 1093-1104, 2023.

2022:

Vergara, L. P., Hackbart, H. C. dos S., Alves, C. J., Reissig, G. N., Wachholz, B. S., Borges, C. D., Chim, J. F., Zambiazi, R. C. Encapsulation of phenolic compounds through the complex coacervation technique for the enrichment of diet chewable candies. *Food Bioscience*, v. x, p. 102256, 2022.

Reissig, G. N., Fajardo, Y. C. G., Parise, A. G., Ribeiro, R. V., Souza, G. M. Experimental Evidence for Fruit Memory and Its Applications to Post-harvest Physiology and Technology: An Overview. In: Ulrich Lüttge; Francisco M. Cánovas; María-Carmen Risueño; Christoph Leuschner. (Org.). *Progress in Botany*. 1ed. Berlin: Springer, 2022, v. 83, p. 1-24.

2021:

Parise, A. G., Reissig, G. N., Basso, L. F., Senko, L. G. S., Oliveira, T. F. C., De Toledo, G. R. A., Ferreira, A. S., Souza, G. M. Detection of Different Hosts From a Distance Alters the Behaviour and Bioelectrical Activity of Cuscuta racemosa. *Frontiers in Plant Science,* v. 12, p. 1-21, 2021.

Reissig, G. N, Oliveira, T. F. C., Oliveira, R. P., Posso, D. A., Parise, A. G., Nava, D. E., Souza, G. M. Fruit Herbivory Alters Plant Electrome: Evidence for Fruit-Shoot Long-Distance Electrical Signaling in Tomato Plants. *Frontiers in Sustainable Food Systems,* v. 5, p. 657401, 2021.

Reissig, G. N., Oliveira, T. F. C., Costa, Á. V. L., Parise, A. G., Pereira, D. R., Souza, G. M. Machine Learning for Automatic Classification of Tomato Ripening Stages Using Electrophysiological Recordings. *Frontiers In Sustainable Food Systems*, v. 5, p. 696829, 2021.

Vergara, L. P., Reissig, G. N., Zambiazi, R. C., Rodrigues, R. da S., Chim, J. F. Red pitanga chewable candies: physicochemical, microbiological, and sensory characterization. *Ciência E Tecnologia De Alimentos (Online)*, v. x, p. Ahead of Print, 2021.

Parise, A. G., Basso, L. F., Oliveira, R. P., Reissig, G. N. Localised photosynthetic pigments in the node of a holoparasitic plant: support for shoot growth?. *Acta Botanica Brasilica,* v. 35, p. 698-702, 2021.

Reissig, G. N., Parise, A. G., Posso, D. A., Costa, A. V. L., Vergara, L. P., Oliveira, T. F. C., Rombaldi, C. V. From the Plant to the Body:

Phytochemicals and their Bioactivity for Humans. In: Gulcin Yildiz; Gökçen İzli. (Org.). *Bioactive Compounds and their Importance.* 1ed. Hauppauge: Nova Science Publishers, Inc., 2021, v., p. 163-.

Thiago Francisco de Carvalho Oliveira

Affiliation:
Federal University of Pelotas (UFPel). Laboratory of Cognition and Plant Electrophysiology (LACEV)

Education:
Bachelor's degree in Physics (UFPel). Master's in Plant Physiology (UFPel). PhD candidate in Plant Physiology (UFPel).

Business Address:
Federal University of Pelotas, Department of Botany. University Campus, Andre Dreyfuss Alley building 21. Jardim América. ZIP code: 96160000 - Capão do Leão, RS – Brazil

Research and Professional Experience:
Specialist in plant electrophysiology, physics of chaotic systems, mathematics in stochastic and complex signal processing, plant physiology and development of analytical systems based on machine learning.

Publications from the Last 3 Years:

2023:

Costa, Á. V. L., Oliveira, T. F. C., Posso, D. A., Reissig, G. N., Parise, A. G., Barros, W. S., Souza, G. M. Systemic Signals Induced by Single and Combined Abiotic Stimuli in Common Bean Plants. Plants, v. 12, p. 924, 2023.

Reissig, G. N., Oliveira, T. F. C., Parise, A. G., Costa, Á. V. L., Posso, D. A., Rombaldi, C. V., Souza, G. M. Approximate entropy: a promising tool to understand the hidden electrical activity of fruit. Communicative & Integrative Biology, v. 16, p. 2195236, 2023.

Parise, A. G., Oliveira, T. F. C., Debono, M-W., Souza, G. M. The Electrome of a Parasitic Plant in a Putative State of Attention Increases the Energy of Low Band Frequency Waves: A Comparative Study with Neural Systems. *Plants,* v. 12, p. 2005, 2023.

2022:

Parise, A. G., De Toledo, G. R. A., Oliveira, T. F. C., Souza, G. M., Castiello, U., Gagliano, M., Marder, M. Do plants pay attention? A possible phenomenological-empirical approach. *Progress In Biophysics & Molecular Biology,* v. 173, p. 11-23, 2022.

2021:

Parise, A. G., Reissig, G. N., Basso, L. F., Senko, L. G. S., Oliveira, T. F. C., De Toledo, G. R. A., Ferreira, A. S., Souza, G. M. Detection of Different Hosts From a Distance Alters the Behaviour and Bioelectrical Activity of Cuscuta racemosa. Frontiers in Plant Science, v. 12, p. 1-21, 2021.

Reissig, G. N, Oliveira, T. F. C., Oliveira, R. P., Posso, D. A., Parise, A. G., Nava, D. E., Souza, G. M. Fruit Herbivory Alters Plant Electrome: Evidence for Fruit-Shoot Long-Distance Electrical Signaling in Tomato Plants. Frontiers in Sustainable Food Systems, v. 5, p. 657401, 2021.

Reissig, G. N., Oliveira, T. F. C., Costa, Á. V. L., Parise, A. G., Pereira, D. R., Souza, G. M. Machine Learning for Automatic Classification of Tomato Ripening Stages Using Electrophysiological Recordings. Frontiers In Sustainable Food Systems, v. 5, p. 696829, 2021.

Reissig, G. N., Parise, A. G., Posso, D. A., Costa, A. V. L., Vergara, L. P., Oliveira, T. F. C., Rombaldi, C. V. From the Plant to the Body: Phytochemicals and their Bioactivity for Humans. In: Gulcin Yildiz; Gökçen İzli. (Org.). *Bioactive Compounds and their Importance.* 1ed. Hauppauge: Nova Science Publishers, Inc., 2021, v., p. 163-.

Douglas Antônio Posso

Affiliation:

Federal University of Pelotas (UFPel). Laboratory of Cognition and Plant Electrophysiology (LACEV).

Education:

Bachelor's degree in Agronomy (UFPel). Master's and PhD in Plant Physiology (UFPel).

Business Address:
Federal University of Pelotas, Department of Botany. University Campus, Andre Dreyfuss Alley building 21. Jardim América. ZIP code: 96160000 - Capão do Leão, RS – Brazil.

Research and Professional Experience:
Postdoctoral researcher investigating electrophysiology and cognitive processes in plants. Specialist in biochemical and metabolic analysis. Curious about "how plant's work systematically and in the ecosystems," also likes to teach plant science to others, children and teenagers, specially.

Publications from the Last 3 Years:

2024:
Souza, G. M, Posso, D. A., Oliveira, T. F. C. The quest for time in plant physiology: a processual perspective. *Theoretical And Experimental Plant Physiology*, in press.

2023:
Posso, D. A., Da-Silva, C. J., Shimoia, E. P., Martins, T. da S., Reissig, G. N., De Oliveira, A. C. B., Borella, J., Van Dongen, J. T., do Amarante, L. Root-hypoxia tolerance in soybean sister-lines plants indicates a better balance in energy use/dissipation and oxidative stress control. *Plant Stress,* v. 10, p. 100225, 2023.

Costa, Á. V. L., Oliveira, T. F. C., Posso, D. A., Reissig, G. N., Parise, A. G., Barros, W. S., Souza, G. M. Systemic Signals Induced by Single and Combined Abiotic Stimuli in Common Bean Plants. *Plants*, v. 12, p. 924, 2023.

Martins, T. da S., Da-Silva, C. J., Shimoia, E. P., Posso, D. A., Carvalho, I. R., De Oliveira, A. C. B., Do Amarante, L. Nitrate supply decreases fermentation and alleviates oxidative and ionic stress in nitrogen-fixing soybean exposed to saline waterlogging. *Functional Plant Biology*, v. 50, p. 416-433, 2023.

Reissig, G. N., Oliveira, T. F. C., Parise, A. G., Costa, Á. V. L., Posso, D. A., Rombaldi, C. V., Souza, G. M. Approximate entropy: a promising tool to understand the hidden electrical activity of fruit. *Communicative & Integrative Biology*, v. 16, p. 2195236, 2023.

Shimoia, E. P., Da-Silva, C. J., Posso, D. A., Da Silva Martins, T., Agualongo, D. A. P., De Oliveira, A. C. B., Do Amarante, L. Co-inoculation of Seeds with Bradyrhizobium, Azospirillum, and Rhizophagus Improves Nitrogen Assimilation and Growth in Soybean Plants Subjected to Waterlogging. *Russian Journal Of Plant Physiology,* v. 70, p. 146, 2023.

2022:

Hüther, C. M., Borella, J., Martinazzo, E. G., Dalberto, D. S., De Farias, M. E., Posso, D. A., Rombaldi, C. V, Bacarin, M. A. Mitochondrial sense sHSP23.6 protein keeps photosynthetic electron transport during drought. *Theoretical And Experimental Plant Physiology,* v. 34, p. 95-108, 2022.

Agualongo, D. A. P., Da-Silva, C. J., Garcia, N., De Oliveira, F. K., Shimoia, E. P., Posso, D. A., De Oliveira, A. C. B., De Oliveira, D. dos S. Colares, Do Amarante, L. Waterlogging priming alleviates the oxidative damage, carbohydrate consumption, and yield loss in soybean (Glycine max) plants exposed to waterlogging. *Functional Plant Biology*, v. 49, p. FP22030, 2022.

2021:

Reissig, G. N., Oliveira, T. F. C., Oliveira, R. P., Posso, D. A., Parise, A. G., Nava, D. E., Souza, G. M. Fruit Herbivory Alters Plant Electrome: Evidence for Fruit-Shoot Long-Distance Electrical Signaling in Tomato Plants. *Frontiers In Sustainable Food Systems*, v. 5, p. 244, 2021.

Da-Silva, C. J., Shimoia, E. P., Posso, D. A., Cardoso, A. A., Batz, T. A., Oliveira, A. C. B., Do Amarante, L. Nitrate nutrition increases foliar levels of nitric oxide and waterlogging tolerance in soybean. *Acta Physiologiae Plantarum*, v. 43, p. 116, 2021.

Reissig, G. N., Parise, A. G., Posso, D. A., Costa, A. V. L., Vergara, L. P., Oliveira, T. F. C., Rombaldi, C. V. From the Plant to the Body: Phytochemicals and their Bioactivity for Humans. In: Gulcin Yildiz; Gökçen İzli. (Org.). *Bioactive Compounds and their Importance*. 1ed. Hauppauge: Nova Science Publishers, Inc., 2021, v., p. 163-.

Helena Chaves Tasca

Affiliation:

Federal University of Pelotas (UFPel). Laboratory of Cognition and Plant Electrophysiology (LACEV).

Education:
Licensed in Biological Sciences (URI). Master's in Ecology (URI). PhD candidate in Plant Physiology (UFPel).

Business Address:
Federal University of Pelotas, Department of Botany. University Campus, Andre Dreyfuss Alley building 21. Jardim América. ZIP code: 96160000 - Capão do Leão, RS – Brazil\

Research and Professional Experience:
Experience in Ecology, Plant Physiology, Plant Stress (water and temperature), Education and Scientific Communication

Lyana Pintos Ramos

Affiliation:
Federal University of Pelotas (UFPel). Laboratory of Cognition and Plant Electrophysiology (LACEV).

Education:
Undergraduate student in Biological Sciences Bachelor's program (UFPel).

Business Address:
Federal University of Pelotas, Department of Botany. University Campus, Andre Dreyfuss Alley building 21. Jardim América. ZIP code: 96160000 - Capão do Leão, RS – Brazil.

Research and Professional Experience:
Participates in the development of research related to plant intelligence, cognitive aspects, mainly plant-plant communication and electrophysiology, analyzing plant responses in relation to abiotic stress.

Publications from the Last 3 Years:
Reissig, G. N., Basso, L. F., Ramos, L. P., de Souza, J. G. M., de Almeida, L. B. Inteligência E Eletrofisiologia Vegetal. *Expressa Extensão* (UFPEL), v. 28, p. 61-66, 2023.

João Gabriel Moreira de Souza

Affiliation:
Federal University of Pelotas (UFPel). Laboratory of Cognition and Plant Electrophysiology (LACEV).

Education:
Undergraduate student in Biological Sciences Bachelor's program (UFPel).

Business Address:
Federal University of Pelotas, Department of Botany. University Campus, Andre Dreyfuss Alley building 21. Jardim América. ZIP code: 96160000 - Capão do Leão, RS – Brazil

Research and Professional Experience:
Studying plant physiology, specifically electrophysiology as means of understanding plant consciousness and intelligence.

Publications from the Last 3 Years:
Reissig, G. N., Basso, L. F., Ramos, L. P., de Souza, J. G. M., de Almeida, L. B. Inteligência E Eletrofisiologia Vegetal. *Expressa Extensão* (UFPEL), v. 28, p. 61-66, 2023.

Luis Felipe Basso

Affiliation:
Federal University of Pelotas (UFPel). Laboratory of Cognition and Plant Electrophysiology (LACEV).

Education: Bachelor's degree in Biological Sciences (UFPel).

Business Address:
Federal University of Pelotas, Department of Botany. University Campus, Andre Dreyfuss Alley building 21. Jardim América. ZIP code: 96160000 - Capão do Leão, RS – Brazil.

Research and Professional Experience:
Plant Ecophysiology and Ecological interactions in plants.

Publications from the Last 3 Years:

2023:
Reissig, G. N., Basso, L. F., Ramos, L. P., de Souza, J. G. M., de Almeida, L. B. Inteligência E Eletrofisiologia Vegetal. *Expressa Extensão* (UFPEL), v. 28, p. 61-66, 2023.

2021:
Parise, A. G., Reissig, G. N., Basso, L. F., Senko, L. G. S., Oliveira, T. F. C., De Toledo, G. R. A., Ferreira, A. S., Souza, G. M. Detection Of Different Hosts From A Distance Alters The Behaviour And Bioelectrical Activity Of Cuscuta Racemose, *Frontiers In Plant Science,* V. 12, P. 594195, 2021.
Parise, A. G., Basso, L. F., Oliveira, R. P., Reissig, G. N. Localised Photosynthetic Pigments In The Node Of A Holoparasitic Plant: Support For Shoot Growth? *Acta Botanica Brasilica*, V. 35, P. 698-702, 2021.

André Geremia Parise

Affiliation:
School of Biological Sciences, University of Reading, Reading, UK.

Education:
Bachelor's degree in Biological Sciences (Federal University of Santa Catarina (UFSC), Brazil). Master's degree in plant Physiology (Federal University of Pelotas (UFPel), Brazil).

Business Address:
Health and Life Sciences building, School of Biological Sciences, University of Reading, Whiteknights Campus, RG6 6EX, Reading, UK.

Research and Professional Experience:
PhD Student at the University of Reading. Studies non-neural cognition and ecophysiology in plants and, more recently, fungi as well.

Publications from the Last 3 Years:

2023:

Parise A. G., Marder M. (2023). Extended Plant Cognition: a critical consideration of the concept. *Theoretical and Experimental Plant Physiology*. doi: 10.1007/s40626-023-00281-5.

Parise A. G., Oliveira T. F. C., Debono M. W., Souza G. M. (2023). The electrome of a parasitic plant in a putative state of attention increases the energy of low band frequency waves: a comparative study with neural systems. *Plants,* 12(10): 2005. doi: 10.3390/plants12102005.

Reissig G. N., Oliveira T. F. C., Parise A. G., Costa A. V. L., Posso D. A., Rombaldi C. V., Souza G. M. (2023). Approximate Entropy: a promising tool to understand the hidden electrical activity of fruit. *Communicative & Integrative Biology*, 16(1): 2195236. doi: 10.1080/19420889.2023. 2195236.

Costa A. V. L., Oliveira T. F. C., Posso D. A., Reissig G. N., Parise A. G., Barros WS, Souza GM. (2023). Systemic signals induced by single and combined abiotic stimuli in common bean plants. *Plants*, 12(4): 924. doi: 10.3390/plants12040924.

Parise A. G., Gubert G. F., Whalan S., Gagliano M. (2023). Ariadne's thread and the extension of cognition: a common but overlooked phenomenon in nature? *Frontiers in Ecology and Evolution*, 10: 1069349. doi: 10.3389/fevo.2022.1069349.

2022:

Parise A. G., de Toledo G. R. A., Oliveira T. F. C., Souza G. M., Castiello U., Gagliano M., Marder M. (2022). Do plants pay attention? A possible phenomenological-empirical approach. *Progress in Biophysics and Molecular Biology*, 173: 11-23. doi: 10.1016/j.pbiomolbio.2022.05.008.

Reissig G. N., Oliveira T. F. C., Costa A. V. L., Parise A. G., Pereira D. R., Souza G. M. (2021). Machine learning for automatic classification of tomato ripening stages using electrophysiological recordings. *Frontiers in Sustainable Food Systems*, 5: 696829. doi: 10.3389/fsufs.2021.696829.

2021:

Reissig G. N., Oliveira T. F. C., Oliveira R. P., Posso D. A., Parise A. G., Nava D. E., Souza G. M. (2021). Fruit herbivory alters plant electrome: evidence for fruit-shoot long-distance electrical signaling in tomato plants. *Frontiers in Sustainable Food Systems*, 5: 657401. doi: 10.3389/fsufs.2021.657401.

Parise A. G., Bertoli S. C., Souza G. M. (2021). Belowground interactions affect shoot growth in Eucalyptus urophylla under restrictive conditions. *Plant Signaling & Behavior*, 16(9): 1927589. doi: 10.1080/15592324.2021.1927589.

Parise A. G., Basso L. F., de Oliveira R. P., Reissig G. N. (2021). Localised photosynthetic pigments in the node of a holoparasitic plant: support for shoot growth? *Acta Botanica Brasilica*, 35(4): 698-702. doi: 10.1590/0102-33062020abb0302.

Parise A. G., Reissig G. N., Basso L. F., Senko L. G. S., Oliveira T. F. C., de Toledo G. R. A., Ferreira A. S., Souza G. M. (2021). Detection of different hosts from a distance alters the behaviour and bioelectrical activity of Cuscuta racemosa. *Frontiers in Plant Science*, 12: 594195. doi: 10.3389/fpls.2021.594195.

Dori Edson Nava

Affiliation: Embrapa Clima Temperado.

Education: PhD in entomology from the University of São Paulo.

Business Address:
BR 392 Km 78. Mailbox 403. Zip Code: 96010-971, Pelotas, RS, Brazil.

Research and Professional Experience:
Researcher at Embrapa Temperate Climate, where I carry out research on bioecology and biological control of insect pests, especially pests from temperate fruit crops, including fruit flies. The focus of the research is related to the use of biological control in integrated pest management programs.

Professional Appointments:
Researcher at Embrapa Temperate Climate and works as a permanent professor in the Postgraduate courses in Plant Health and Animal Biodiversity at the Federal University of Pelotas.

Publications from the Last 3 Years:

2023:

Zucchi, R. A., Malavasi, A., Adaime, R., Nava, D. E. Mosca-das-frutas no Brasil: conhecimento básico e aplicado - Volume 1. 1. ed. *Piracicaba:* FEALQ, 2023. 549 p.

Zucchi, R. A., Malavasi, A., Adaime, R., Nava, D. E. Mosca-das-frutas no Brasil: conhecimento básico e aplicado - Volume 2. 1. ed. *Piracicaba:* FEALQ, 2023. 397 p.

Aguilera, G. R. V., Scheunemann, T., Kruger, A. P., Correa, L. M. S., Bernardi, D., Nava, D. E. Selection of strains of *Trichogramma foersteri* and *Trichogramma pretiosum* (Hymenoptera: Trichogrammatidae) for the control of *Palpita forficifera* (Lepidoptera: Crambidae). *Neotropical Entomology*, v. 51, p. se, 2023.

Araujo, A. S., Zucchi, R. A., Norrbom, A. L., Nanini, F., Correa, A. S., Alvarenga, C. D., Souza-Filho, M. F., Nava, D. E., Savaris, M. Integrative approach reveals the identity of Brazilian specimens previously recognized as *Anastrepha dissimilis* Stone, 1942 (Diptera: Tephritidae). *Zootaxa,* v. 5228, p. 317-336, 2023.

Molter, A., Bezerra, J. I. M., Rafikova, E., Nava, D. E., Rafikov, M. Dynamics and biological control of the sugarcane borer with two parasitoids. *Ecological Modelling, v.* 428, p. 110371, 2023.

Garcez, A. M., Krüger, A. P., Nava, D. E. Intrinsic competition between 2 pupal parasitoids of *Drosophila suzukii* (Diptera: Drosophilidae). *Annals of the Entomological Society of America,* v. 116, p. 145-153, 2023.

Piovesan, B., Contreras M. J. A., Gonçalves, R. S., Daniel N. S., Bernardi, D., Botton, M., Nava, D. E. Lethal and sublethal effects of toxic bait formulations on *Doryctobracon areolatus* (Hymenoptera: Braconidae) and implications for integrated fruit fly management. *Journal of Economic Entomology*, v.116, p. 855-863, 2023.

Krüger, A. P., Garcez, A. M., Scheunemann, T., Bernardi, D., Nava, D. E., Garcia, F. R. M. Reproductive biology of *Trichopria anastrephae* (Hymenoptera: Diapriidae), a biological control agent of *Drosophila suzukii* (Matsumura) (Diptera: Drosophilidae). *Neotropical Entomology*, v. 52, p. 1-8, 2023.

2022:

Dias, N. P., Montoya, P., Nava, D. E. Historical invasion of Medfly in the Neotropical region and adoption of management techniques. *Current Opinion in Insect Science,* v. 50, p. 100872, 2022.

Dias, N. P., Montoya, P., Nava, D. E. A 30¿year systematic review reveals success in tephritid fruit fly biological control research. *Entomologia Experimentalis et Applicata,* v. 170, p. 1-15, 2022.

2021:

Reissig, G. N., Oliveira, T. F. C., Oliveira, R. P., Posso, D. A., Parise, A. G., Nava, D. E., Souza, G. M. Fruit Herbivory alters plant electrome: evidence for fruit-shoot long-distance electrical signaling in tomato plants. *Frontiers in Sustainable Food Systems*, v. 5, p. 5:657401., 2021.

Sousa, M. S. M., Santos, J. E. V., Toledo, J. J., Nava, D. E., Adaime, R. *Geissospermum argenteum* (Angiosperma: Apocynaceae): a reservoir of parasitoids of fruit flies (Diptera: Tephritidae) in an upland forest in the Brazilian Amazon. *Agricultural and Forest Entomology,* v. 23, p. 297-310, 2021.

Parra, J. R. P., Pinto, A. S., Nava, D. E., Oliveira, R. C., Diniz, A. J. F. Controle biológico com parasitoide e predadores na agricultura brasileira. 1. ed. *Piracicaba: Fealq,* 2021. v. 1. 592p

Chapter 7

Comparison of the Basic Chromosome Topologies of Onion (*Allium cepa*) from Germany and China

Kristina Zubow[1,*]
Anatolij Zubow[2]
and Viktor A. Zubow[1]
[1]R&D, Zubow Consulting (NPO), Waren, Germany
[2]Dept. Telecommunication Networks Group, TU Berlin, Berlin, Germany

Abstract

The basic topology of the molecular architecture of the chromosomes of the onion *Allium cepa* from China and northern Germany was studied using the gravitational mass spectroscopy (GMS) method. Signals from nucleosomes (with varying degrees of DNA winding) in which loose conformations dominate were detected. Nucleosomes are formed into basic solenoid structures with turns of six histones. These solenoids-I are further twisted into the second level of chromosome topology - solenoids-II. Solenoids-II are found only in Germany onions. Circular mitDNAs (~11 MDa) are detected in both onion varieties, and large ones (~99 MDa) only in onions from Germany. The distribution of nucleosomes, solenoids and mitDNA in different tissues of *Allium cepa* (basal plate, lateral and apical buds) was studied.

Keywords: *Allium cepa*, nucleosomes, solenoids, chromosomes, mitDNA, distribution in onions, long-range order

---

* Corresponding Author's Email: vzubow@googlemail.com.

In: Advances in Biology. Volume 6
Editor: Charles D. Grant
ISBN: 979-8-89113-712-7

1. Introduction

In recent years, the attention of a large group of scientists has been focused on studying the onion genome [1, 2]. It is reported that only about 70% of the genome could be identified [3]. However, the most important aspect, the basic topology of chromosomes, is not given enough attention due to the lack of a non-destructive testing method, also regarding mitochondrial DNA (mitDNA) [4]. Although everyone understands that their conformations in the native state play a decisive role in the correct understanding of the mechanism of energetic interaction of onions with the environment, supplying energy to neurons and then chromosomes. In chromosomes, nucleosomes are the elementary cells of information, and their conformations, like piezo electrics [9], are highly dependent on the energy supply of mitDNA. Therefore, special attention had to be paid to the forms of mitDNA. Thus, in [4] it is reported that for *Alliumcepa L.* circular mitDNA have a mass equivalent to 316,363 bp (https://www.ncbi.nlm.nih.gov/nuccore/NC_030100.1), and for the wild variety only 154,804 bp (https://www.tandfonline.com/doi/full/10.1080/23802359.2019.1572462), while the reasons are not discussed, and the role of conformations is generally omitted.

Since chromosomes control all processes in a cell, including activation/deactivation of genes, that is, they are the brain of cells, and in the ensemble of chromosomes of plant cells, in general, they are the brain of an onion, its neural network, then a targeted study of the state of at least the basic topology of the chromosomes of this plant is necessary.

The purpose of this article was to study the state of the basic chromosome topology of European and Asian varieties of edible onions (*Allium cepa*), the masses and conformations of medium and large mitDNA.

2. Material and Method

The object of the study was the young onion of the *Allium cepa* variety, cultivated in 2023 in the north of Germany (Upper Pomerania) and the *Allium cepa* variety, supplied from China. Gravitational mass spectroscopy (GMS) was used as a non-destructive diagnostic of the cytoskeleton and long-range molecular order in onions. The GMS sensor was inserted into the vegetable (basal plate, lateral and apical bud of the onion) to a depth of 0.1 mm, and gravitational noise (GN) from oscillations of atomic nuclei clusters (ANC) was recorded in the long-range order (LRO) of the plant cytoskeleton (ANC, clusters, domains), coils, micellar and supra micellar formations, genes, etc.). GMS spectra were obtained using the first Zubow equation [5]. Positive values of Δf are characteristic of loose conformations of ANC; they reflect the energy part of the molecular cluster in the entire ensemble of studied mass concentrations; negative values $-\Delta f$ reflect the same thing, but for dense conformations of ANC. Δf is the result of subtracting the natural background from the spectrum of gravitational noise (GN) in the GN sample. The samples were protected as much as possible from external energetic physical influences (electromagnetic radiation, mechanical fields and strong sudden pressure changes). Figure 1 shows a schematic plan of the method.

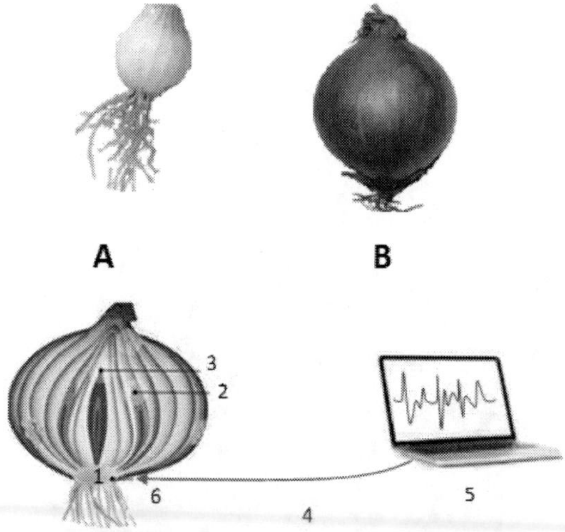

Figure 1. Samples of *Allium cepa* onion, origin: **A** – northern Germany and **B** – China. Schematic plan of the analysis method. 1 – basal plate (https://en.wikipedia.org/wiki/Bulb), 2 - lateral bud, 3 - apical bud, 4 - interface, 5 - computer and 6 - sensor.

3. Results and Discussion

Figure 2 shows a high-resolution photograph of a plant chromosome and a model for the formation of solenoid structures in it in the basic topology with tense bends of the solenoid, and Figure 3 shows a model for the formation of the first solenoid-I structure in the basic topology of chromosomes, for understanding.

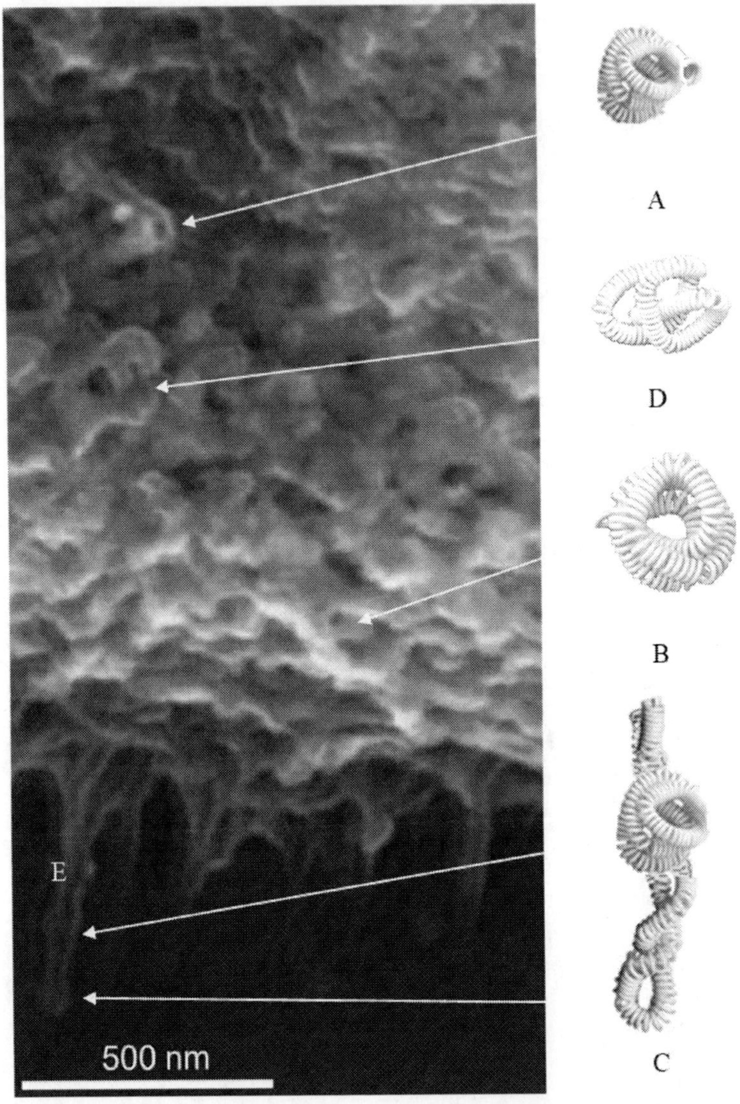

Figure 2. Fragment of a photograph of a chromosome [6] and a model of its globular construction from folded solenoids of nucleosomes [7]. **A** - a model of a micro globule with end solenoids ($Ø_{out} \approx 48$ nm, $Ø_{int} \approx 24$ nm), **B** - toroidal micro globule ($Ø_{out} \approx 120$ nm, $Ø_{int} \approx 24$ nm) made of solenoids. **C** - model of terminal micro globule ($Ø \approx 71$ nm, calculation: 89,743,983 Da, found 89,816,036 Da, mostly dense conformations, error 0.080%), **Z** = 6.166E-15 N/m. **D** - unfolded spiral. **E** - knots in twisted spirals to minimize potential energy [8]. On the right, for understanding, models of the formation of the solenoid structure of the basic chromosome topology are given.

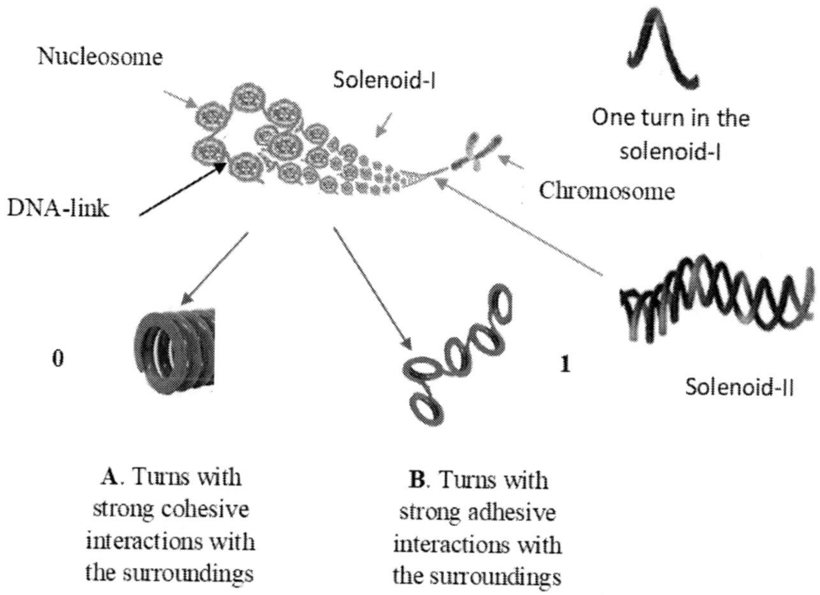

Figure 3. Model of two types of states of turns in a solenoid-I of histones-6 with strong (highly compressed solenoid) and weak cohesive interactions with surrounding analogs (highly stretched solenoid). There are 6 nucleosomes in one turn (top). In **A**, the octamer in the nucleosome is not compressed, solenoid-I (**0**) without intra-turn stresses and in **B**, on the contrary, internal stresses on the bends (**1**) weaken the chemical bonds in the DNA-links. https://yandex.com/images/search?img_url= https%3A%2F%2Fkonspekta.net%2Flektsianew%2Fbaza9%2F510345814039.files %2Fimage078.jpg&lr=112251&pos=3&rpt=simage&source=serp&stype=image&text==topology%20chromewasp%20pictures.

Figure 4 shows the GMS spectra of an ensemble of nucleosomes (https://de.wikipedia.org/wiki/Nukleosom) in the basic chromosome topology of *Allium cepa* (Germany). The Figure shows that the number of types of nucleosomes in the basal plate is greater than in the apical bud by two kinds, which are indicated by an asterisk. In general, in both parts of the onion, the dominant conformations of nucleosomes are in loose forms, forms that actively interact with their surroundings. Only such forms of memory cells [9] are capable of controlling all processes in separate cells and ensembles of cells, that is, cells throughout the entire onion. Here, it is worth noting that in chromosomes, nucleosomes are represented predominantly by single-turn octamer coatings with DNA snare loops (log m = 3.6...3.8 Da) or pro nucleosomal structures with incomplete DNA coverage. The fact is that only such structures are active piezo electrics - memory cells. Nucleosomes with a double coating of DNA snare loops are not active in neural systems and plants do not rationally use them [9]. If the dominance of loose conformations of nucleosomes in the basal plate region is obvious, then in the middle of the onion this dominance is weakened, especially for nucleosomes with masses log m = 3.7. The reason for this phenomenon may be a weakening of the resonance influence of the ensemble of nucleosomes in the basal plate on their distant ensembles in the cells of the middle of the onion.

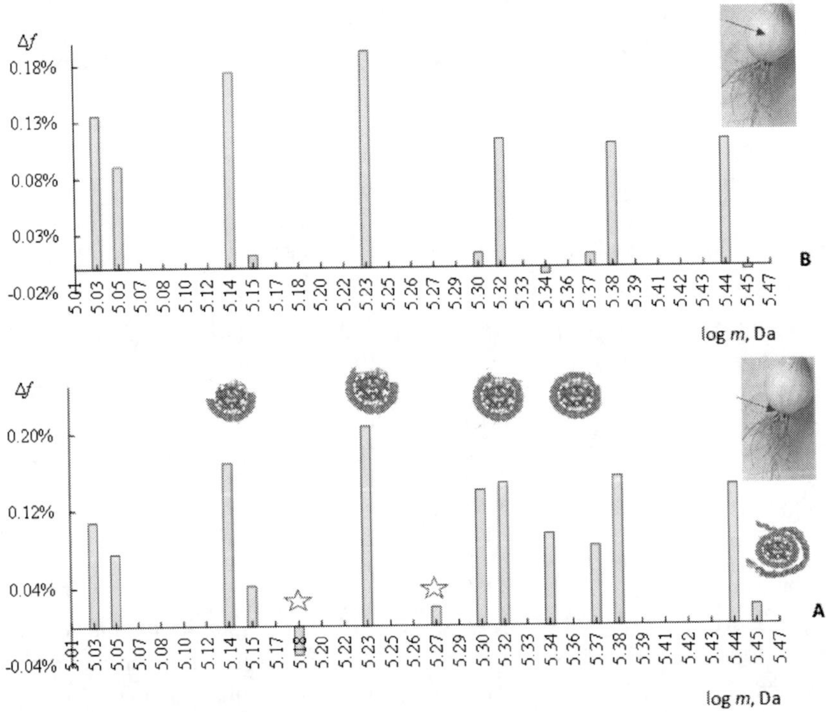

Figure 4. Fragments of GMS spectra of rapidly maturing *Allium cepa* (Germany) in the area of nucleosome oscillations in the basic chromosome topology: **A** – basal plate and **B** – in the middle part of the onion (immersion of the sensor in the middle of the onion).

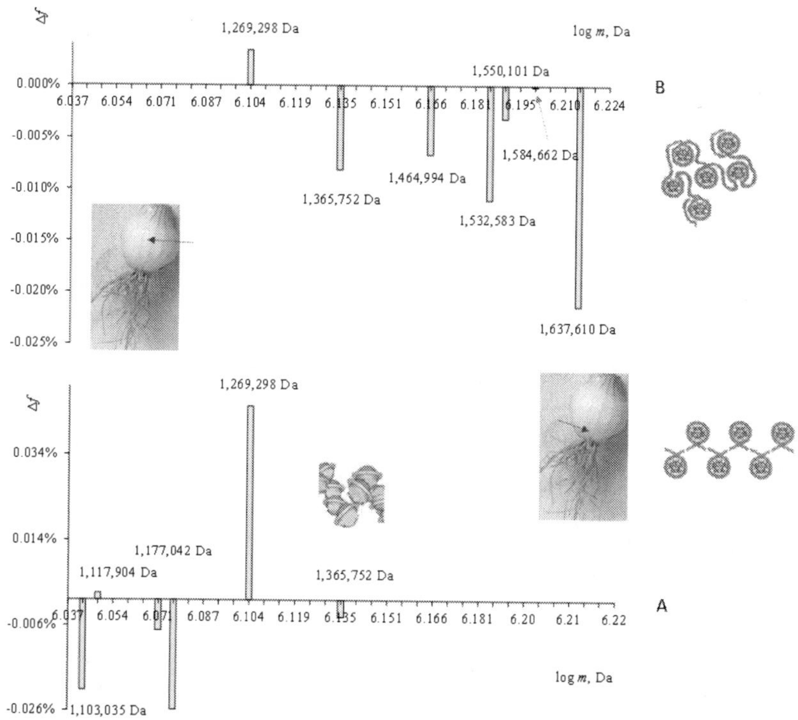

Figure 5. Fragments of GMS spectra of rapidly maturing *Allium cepa* (Germany) in the region of oscillations of one turn of 6 nucleosomes (primary turn) in solenoid-I of the basic chromosome topology: **A** – basal plate and **B** - in the middle part of the onion (immersion of the sensor in the middle of the onion). Models of the state of turns for **A** and **B** are given on the right.

Figure 5 shows fragments of GMS spectra in the region of oscillations of turns of solenoids-I, built from 6 nucleosomes in one turn. Strong differences in the masses of turns are visible. If at the basal plate of the onion from Germany the turns are represented mainly by dense conformations, then in the center they are more massive oscillators and with a clear dominance of dense structures. Here, the common signal is from a coil with a mass of 1,269,298 Da, which in both parts of the onion is presented in loose forms that interact with its surroundings. Higher masses of turns in the center of the onion may indicate an increase in the masses of oscillators due to DNA-links between nucleosomes as a result of the loss of octamers by nucleosomes. These links, in dense structures, are integrated into the turns, like single pendulums. In Figure 5, their models are given on the right for understanding. The following

Figures 5 and 7 show fragments of GMS spectra in the region of oscillations of circular mitDNA in the rapidly ripening onion *Allium cepa* (Germany). Two regions of oscillations of medium mitDNA with masses of ~11 million Da (**A** and **B**) and large ~99 million Da (**C** and **D**) were studied. In the region of mitDNA oscillations (~203 million Da [4]), in both onions, no signals of giant mitDNA were detected. The spectra of medium mitDNA in the region of ~11 million Da (**A** and **B**) are different for different detection locations. Note that the strong signal for 10,643,452 Da in the basal plate of the onion from oscillations of loose ring conformations, in the region of the apical buld (3, Figure 1) is almost extinguished, it is weakly dominated by dense ring shapes (eights and deformed, folded eights). This ANC is similar in mass to human mitDNA (10,653,867 Da or 16569 bp). Apparently, with increasing distance, the resonance field of ensembles of medium mitDNA of the basal plate has a weakened effect on the ensemble of its analogues in the center of the onion, where the concentration of mitDNA is lower.

Let us return to Figures 4 and 5, since the entire spectrum of nucleosomes is represented by their loose conformations, and the spectra of turns in solenoid-I indicate their various forms, we can propose a model for the interaction of turns with each other, through octamer hairpins, https://ru.wikipedia.org/wiki/Hairpin_(biology), Figure 6. According to this model, states of turns (**A**), with dominance of inter turn interactions, should give signals from loose conformations of turns, and state (**B**) from dense ones. The ability of turns to "fix" their conformational states through cooperative interaction with neighboring turns, through the hooks of octamer proteins, may explain the formation of long-term plant memory.

Figure 7 shows the GMS spectra of the mass-average circular mitDNA in German onion. Significant differences in mitDNA states in different parts of the bulb are visible. In the basal plate of the onion, out of 10 signals of oscillating rings, mainly in dense conformations, the signal for loose masses of 10,643,900 Da stands out, while for the middle part of the onion, 8 signals come from dense conformations of mitDNA and only from 2 loose ones, but with slightly larger masses: 10,879,987 and 10,925,828 Da. At the same time, signal 10,643,990, which is closest to human mitDNA (16,569 bp or 10,653,867 Da), is weakly expressed, and there is some dominance of dense ring shapes (collapsed figures of eights). It can be assumed that such closeness is typical for all eukaryotes (https://ru.wikipedia.org/wiki/Eukaryotes). The reasons for differences in mitDNA conformations in cells of different parts of onions can be understood from the standpoint of the main functions of rings in dense conformations, namely the synthesis of sugars [10]. Dense forms of mitDNA in the process of oscillations, under the influence of gravitational noise, (adiabatic changes in conformations) compress the complex of carbon dioxide with water $(H_2CO_3)_6$ to form sugars (for example, to α-D-glucopyranose $C_6H_{12}O_6$).

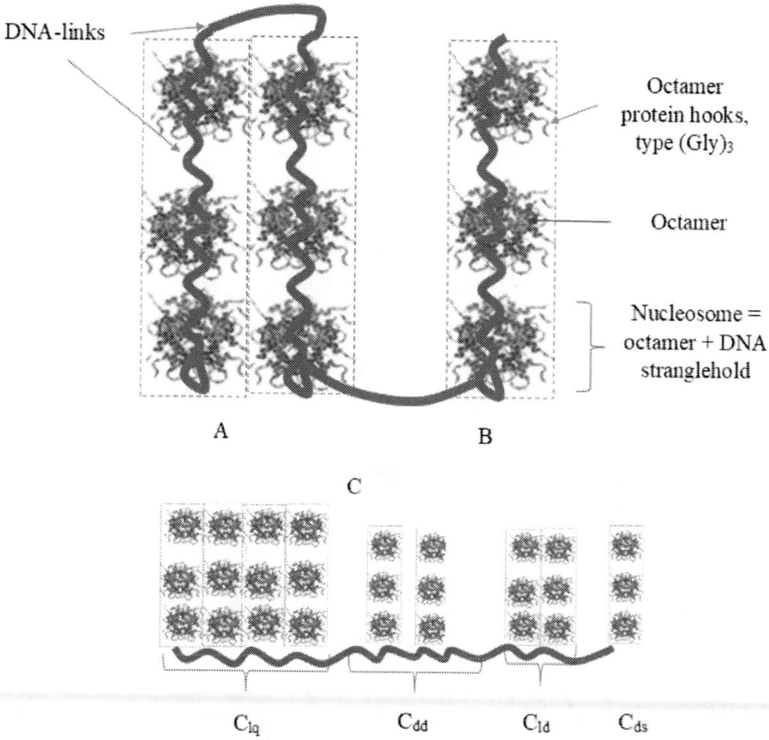

Figure 6. Model of turns of histones-6 (solenoid-I), side view. **A** - 2 turns interact with each other by hooks of fragments of macromolecules of proteins of the octamer, this is a fragment of a dense solenoid-I and **B** - the 3rd turn does not interact with analogs of the surroundings, a fragment of an extended solenoid-I. **A** - loose conformations of turns, as a result of the dominance of interactions with the surroundings in them, **B** - dense conformations of turns (-Δ*f*), as a result of the dominance of intra-turn interactions in them and weakening of interactions with the surroundings. **C** - model of long-term memory formation as fixation by octamer hooks of adjacent turns of nucleosome solenoids in short-term memory conformations. Where are examples: C_{lq} - loose quartet of 4 turns in loose conformations, C_{dd} - dense duo, C_{ld} - loose duo and C_{ds} - dense conformation of one turn, dense solo.

GMS spectra of large mitDNA rings in *Allium cepa* (Germany), in the range of oscillations of the wild variety *Alliumcepa L.*, regardless of the location of detection, practically coincide; the reference mitDNA signal is indicated by an asterisk (99,538,969 Da or 316,363 bp), Figure 8. So, in this variety, large mitDNA rings in dense, folded conformations are stable and reliably represented in different parts of the bulb. Note that no signals from these mitDNAs were detected in the Chinese sample.

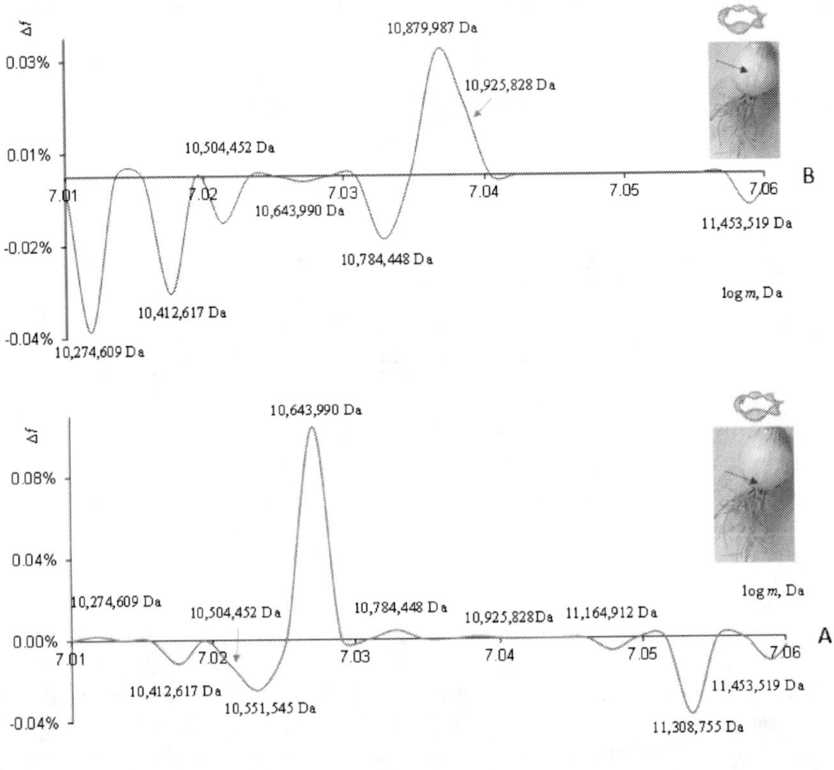

Figure 7. Fragments of GMS spectra in the region of medium mitDNA oscillations in young, rapidly maturing *Allium cepa*, Germany. **A** – basal plate and **B** – in the middle part of the onion (immersion of the sensor in the middle of the onion).

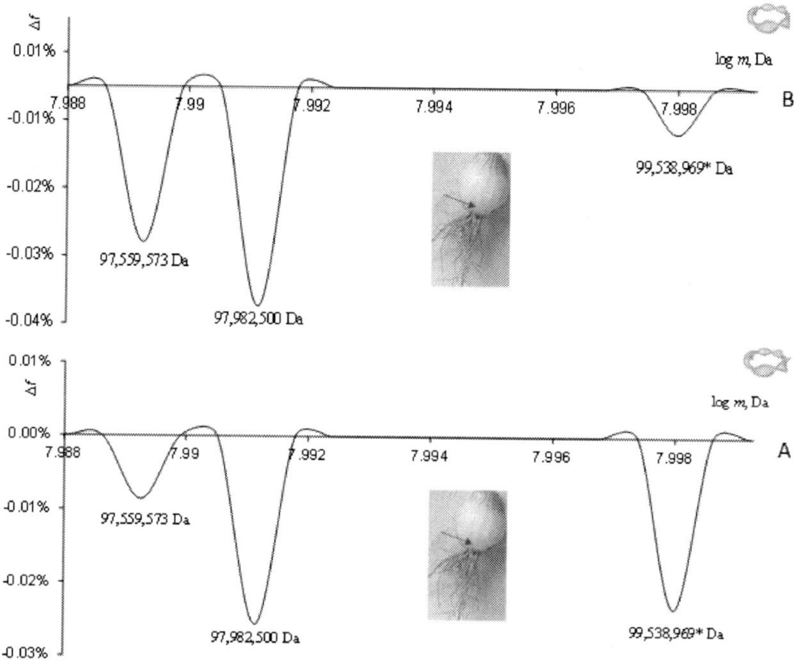

Figure 8. Fragments of GMS spectra in the region of oscillations of large circular mitDNA of the wild variety *Alliumcepa L* [4] in relation to rapidly maturing *Allium cepa*, Germany. **A** – basal plate and **B** – in the middle part of the onion (immersion of the sensor in the middle of the onion).

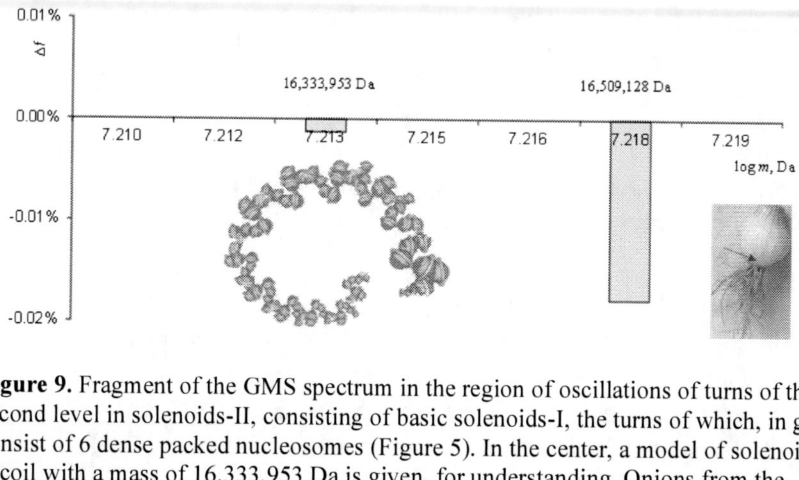

Figure 9. Fragment of the GMS spectrum in the region of oscillations of turns of the second level in solenoids-II, consisting of basic solenoids-I, the turns of which, in go, consist of 6 dense packed nucleosomes (Figure 5). In the center, a model of solenoid-II coil with a mass of 16,333,953 Da is given, for understanding. Onions from the north of Germany.

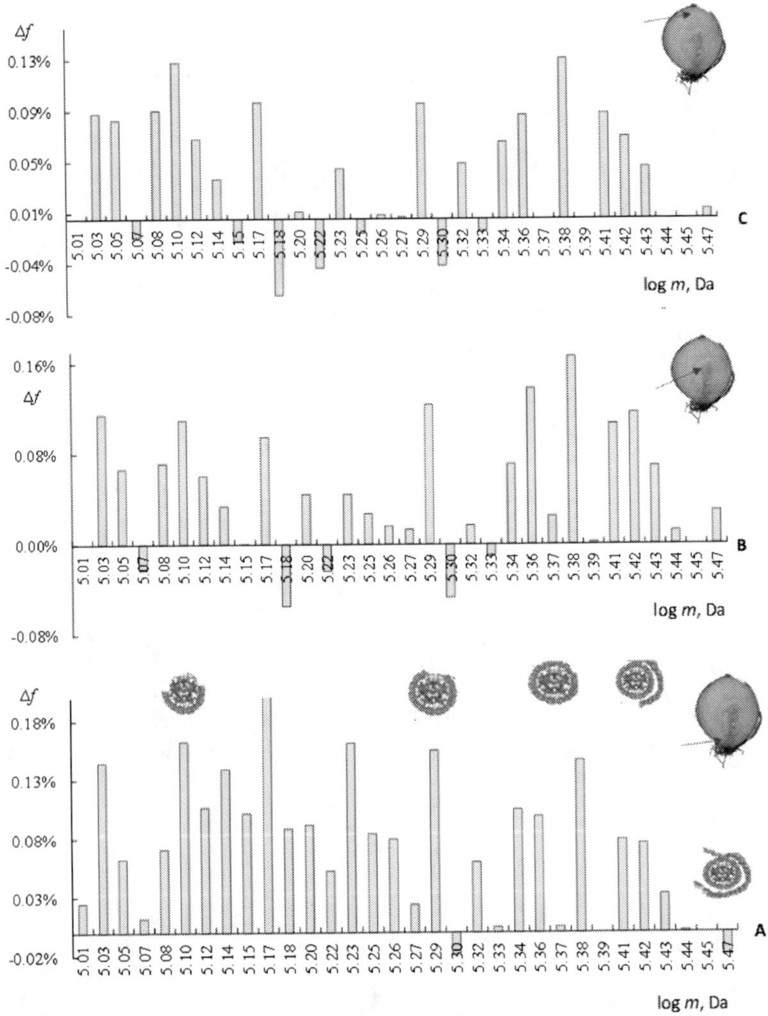

Figure 10. Fragments of GMS spectra of Chinese *Allium cepa* in the area of nucleosome oscillations in the basic topology of chromosomes in: **A** – basal plate, **B** – in the area of the lateral bud and **C** – in the area of the apical bud (see Figure 1).

Figure 10 shows the GMS spectra of nucleosome oscillations in the Chinese onion variety. It can be seen that loose conformations (Δf) dominate in nucleosome ensembles, actively interacting with their surroundings. However, as we move away from the basal plate of the onion, signals of nucleosomes appear in which already dense conformations ($-\Delta f$) dominate. Thus, 2 signals were found in the basal plate, 5 in the lateral bud, and 7 signals from dense forms of nucleosomes in the apical bud. Since dense forms are poor piezo electrics, they do not participate well in the information networks of the onion [9]. The increase in dense forms of nucleosomes with distance from the onion basal plate can be understood from the point of view of the weakening influence of the resonance field from the ensemble of nucleosomes of the basal plate cells on the ensemble of nucleosomes in the apical bud cells. The resonant field of the ensemble of nucleosomes in the basal plate of the onion is weak due to their low concentration, which means that the potential of the basal plate pendulums is not able to fully resonantly control the ensembles of nucleosomes at a distance of 6 cm (the size of the onion). At the same time, the minimization of the potential energy of two ensembles of pendulums identical in nature, as a single oscillator, is weakened and unstable due to the distance and degree of maturation of the onion.

Figure 11 shows the GMS spectrum of Chinese *Allium cepa* in the region of oscillations of one turn of 6 nucleosomes (primary turn) in solenoid-1 of the basic chromosome topology in different parts of the onion. It can notice that the number of signals in the buds is constant and equal to 19, and in the basal plate there is one less. The "extra" signal (expanded ANC, 1,103,035 Da) is marked with an asterisk. The reasons for its appearance require additional research. At the same time, it must be borne in mind that ideal turns of 6 nucleosomes with one turn-noose of DNA on each nucleosome are, with a high probability, rarely represented in solenoids-I. But the combinations, both in terms of the number of nucleosomes in a turn and their degree of DNA winding, are very diverse. That is, there is a lot of imperfection in the topology of onion chromosomes (see Figure 2), which weakens the strength (potential) of the onion bio field as a whole. The bio field of the onion, as a superposition of the resonant fields of all oscillators and their ensembles, was strongly dissipated, and the directionality in the direction of the root was weakened.

Since signals from large (~99 million Da) and giant (~203 million Da) circular mitDNA were not detected in Chinese onion, it was of interest to understand the state of the medium mitDNA in different locations of this onion sample. Figure 12 shows the results of such a study.

As can be seen, the spectra from the buds are very similar (**B** and **C**), but very different from the spectrum (**A**) of mitDNA in the basal plate of the bulb. In the basal plate, mitDNA is predominantly presented in loose forms, while in the buds, on the contrary, in dense conformations. Since the main functions of mitDNA are the synthesis of sugars by mechanochemical compression of a cluster of water with carbon dioxide using the energy of gravitational noise [10], the presence of dense conformations of mitDNA in **B** and **C** becomes clear. Only such ring conformations are able to adiabatically compress the complex $(H_2CO_3)_6$ to the formation of α-D-glucopyranose $C_6H_{12}O_6$ [10].

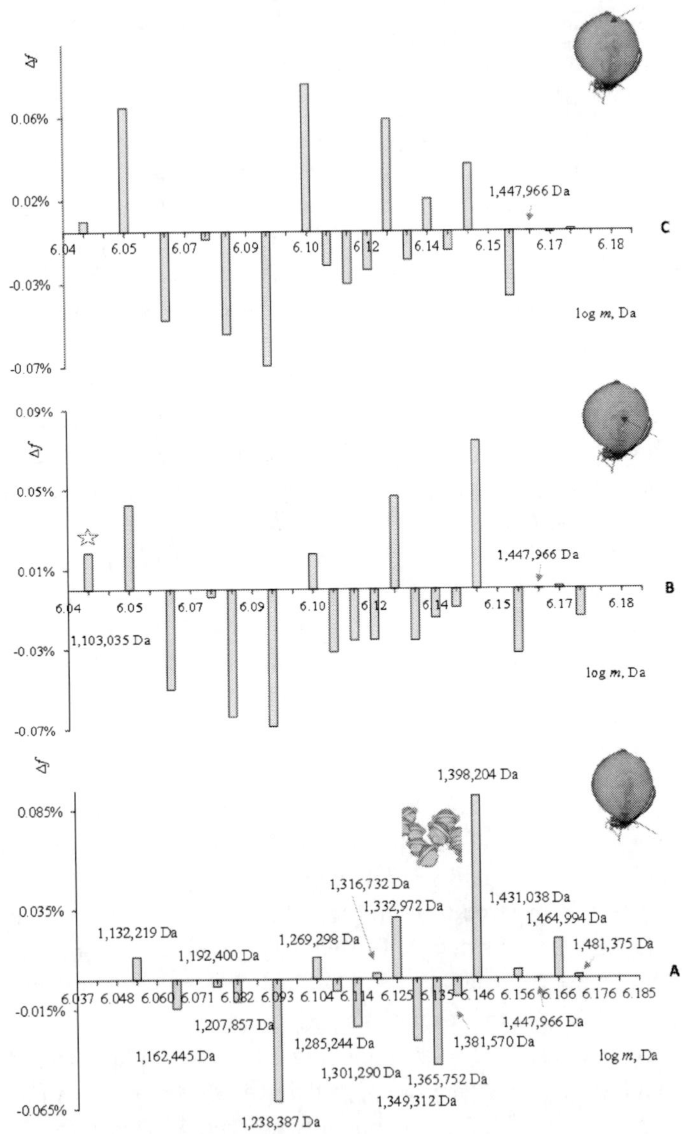

Figure 11. Fragments of the GMS spectra of Chinese *Allium cepa* in the region of oscillations of one turn of 6 nucleosomes (primary turn) in the basic topology of chromosomes, in: **A** - bottom, **B** - in the region of the lateral bud and **C** - in the region of the apical bud (see Figure 1).

In conclusion, we note that in onions from Germany, Ti-plasmid signals from anhydride forms were detected as a result of the use of liquid manure as fertilizer [11], while no such signals were recorded in Chinese onions.

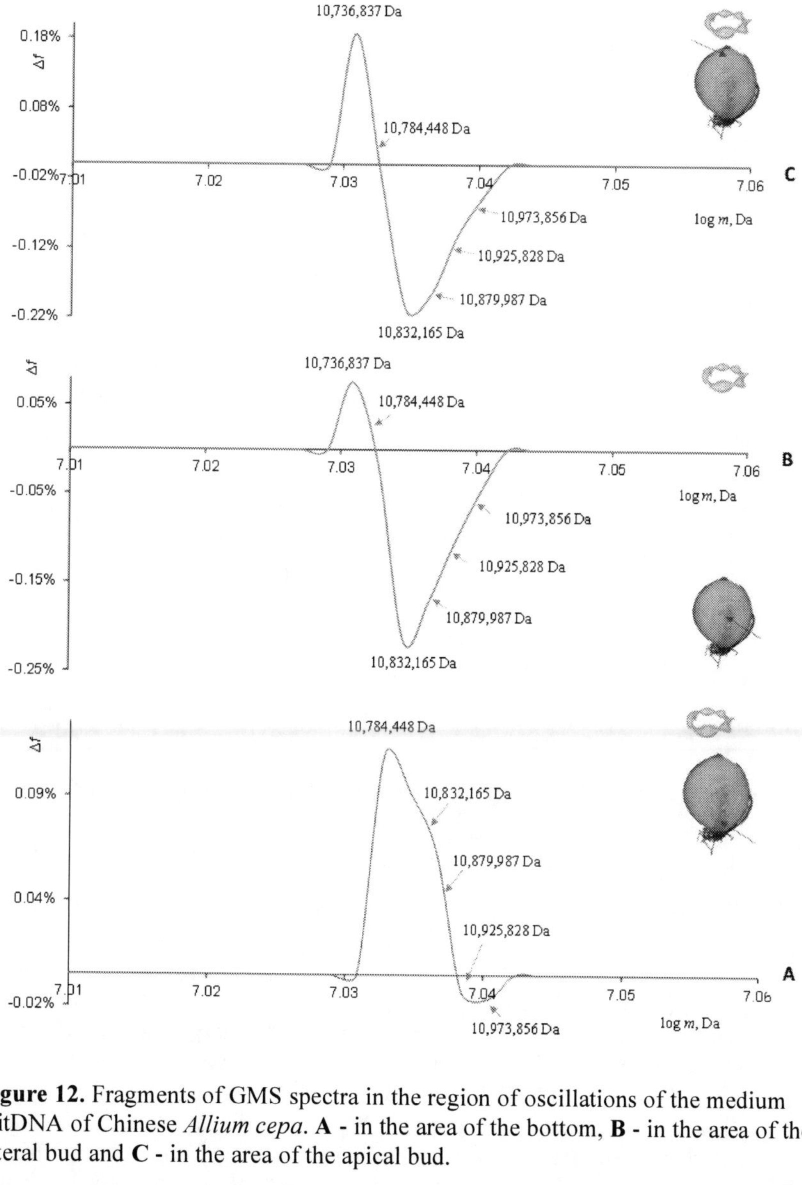

Figure 12. Fragments of GMS spectra in the region of oscillations of the medium mitDNA of Chinese *Allium cepa*. **A** - in the area of the bottom, **B** - in the area of the lateral bud and **C** - in the area of the apical bud.

Conclusion

The GMS method allows you to quickly diagnose the state of the basic topology of chromosomes and mitochondrial DNA of onions.

The GMS method allows you to objectively determine the state of the onion: alive or dead.

The basic topology of chromosomes of Asian and European edible onion varieties is different.

In the European onion variety, the basic chromosome topology is represented by two types of solenoids (I and II), while in the Asian onion there is only one - solenoids-I.

In both varieties, there is a wide range of nucleosomal structures with a pronounced dominance of loose conformations, with strong interaction with the surroundings.

No giant circular mitDNA (~200.3 MDa) was detected in both varieties.

Medium circular mitDNA (~11 MDa) in dense conformations are involved in the synthesis of sugars in the upper parts of the bulbs, and in loose conformations only in information interactions.

The basic topology of chromosomes is resolved and shaped by the gravitational noise of the Universe.

Ensembles of onion cell nucleosomes interact with each other like pendulums through resonant fields.

Sponsoring

To continue work in this direction, we are looking for sponsors. Write to us.

References

[1] Gowd, T. Y. M., Chandra Deo, Dalasanuru Chandregowda Manjunathagowda, Vijay Mahajan, Nangsol Dolma Bhutia Barun Singh. Allelic variability and transferability of atp6 gene among Allium species. *Genetic Resources and Crop Evolution September 2022*, Received: 12 May 2022/Accepted: 12 September 2022, https://doi.org/10.1007/s10722-022-01475-7.

[2] Xinga, J., Mingzhao Zhu, Yongqin Wang and Huiying Liua. The complete mitochondrial genome of Allium fistulosum L. (Amaryllidaceae). *Mitochondrial Dna Part B: Resources 2023*, vol. 8, no. 8, pp. 890–894 https://doi.org/10.1080/23802359.2023.2248684.

[3] Finkers, R., Martijn van Kaauwen, Kai Ament, Karin Burger-Meijer, Raymond Egging, Henk Huits, Linda Kodde, Laurens Kroon, Masayoshi Shigyo, Shusei Sato. Insights from the first genome assembly of Onion (*Allium cepa*). *G3 Genes|Genomes|Genetics*, 2021, Volume 11, Issue 9, September 2021, jkab243, https://doi.org/10.1093/g3journal/jkab243) https://academic.oup.com/g3journal/article/11/9/jkab243/6320787?login=false.

[4] Kim, B., Kim K, Yang T. J., Kim S. Completionof themitochondrial genome sequence of onion (Alliumcepa L.) containing the CMS-S male-sterilecytoplasmand identificationof an independent eventof theccmFNgenesplit. *CurrGenet.* 2016, vol. 62, no. 4, pp. 873–885. doi:10.1007/s00294016-0595-1. https://link.springer.com/article/10.1007/s00294-016-0595-1)

[5] Zubow, K., Zubow A. V., Zubow V. A. The Way to the ETIs. *Applied gravitational mass spectroscopy.* Nova Sci. Publ. NY, 2014.https://www.novapublishers.com/catalog/product_info.php?products_id=42668&osCsid=5bd85d42dc273360fd48126de7be9daf https://www.novapublishers.com/catalog/advanced_search_result.php?keywords=Zubow%20K&sort=2a&page=3&osCsid=59237f459eafe861a2db593a6a504199.

[6] Zoller, J. F. Hochauflösende Strukturanalyse Pflanzlicher Chromosomen In Mitose Und Meiose. *Inaugural-Dissertation zur Erlangung des Doktorgrades der Fakultät für Biologie der Ludwig-Maximilians-Universität München* 2002. https://edoc.ub.uni-muenchen.de/784/1/Zoller_Jutta_F.pdf.

[7] Schroeder-Reiter, E. High resolution analysis of mitotic metaphase chromosomes with scanning electron microscopy: Localizing histone H3 modifications with immunogold labeling in barley (Hordeum vulgare). *Dissertation from the Department of Biology I Electron Microscopy Ludwig-Maximillians-Universität München.* 2004. https://d-nb.info/972840796/34.

[8] Zubow, K., Zubow A., Zubow V. A. Structural Study of Limited Indistant Monitoring of Molecular Domains in the Heart Muscle *in vivo*: Reexamination of the Domains' Mechanicsvariant Sets of Relay Stabilized Systems. In *Horizons in World Cardiovascular Research.* Ed. Eleanor H. Bennington. NY, NovaPublish. Inc. 2017, vol. 13, pp. 227-240. https://www.novapublishers.com/catalog/product_info.php?products_id=62983.

[9] Zubow, A., Zubow K., Zubow V. A. Wegner's phenomenon "I can't think of anything" and the molecular mechanism of mathematical calculations in the brain *in vivo.* Horizons in Neuroscience Research. Eds. Andres Costa & Eugenio Villalba. Nova Publisher Inc. 2020, vol. 43, pp.105…126. *Horizons in Neuroscience Research.* Volume 43 – Nova Science Publishers (novapublishers.com).

[10] Zubow, K., Zubow A., Zubow V. A. Revision of the standard model of photosynthesis. Advances in Chemistry Research. Editor, James C. Taylor. Nova Sci. & Techn. NY, May, 2021, vol. 67, pp. 213-232. *Advances in Chemistry Research.* Volume 67 – Nova Science Publishers (novapublishers.com). ISBN: 978-1-53619-338-1.

[11] Zubow, K., Zubow A., Zubow V. Expansion of R *radiobacter* strain R58. Ti-plasmid – bio time bomb. *Advance in Biology.* 2024. Vol. Ed. Charles D. Grant. Nova Publisher Inc. NY. 2024, in print.

Index

A

Achromobacter, 45, 48
Acidobacteria, 43, 46
Actinobacteria, 43, 44, 45, 46, 48
Actinomyces, 44
active peptides, 38, 55
adjunctive therapy, 38, 55, 57
adjuvant, 53, 59, 110
Aeromonas hydrophila, 56
Allium cepa, vi, vii, 177, 178, 179, 181, 182, 183, 184, 186, 187, 188, 189, 190, 191, 193
allochthonous, 41
amoxicillin, 44, 46, 51, 52, 53, 54, 58
Anhydrides of Ti-bacterial plasmid, 1
antagonistic activity, 56
antibacterial properties, 38
antibiotics, 37, 38, 44, 49, 50, 53, 54, 57
anti-*H. pylori* activity, 56, 57
antimicrobial activity, 55, 57
Atopobium, 44, 46, 47
autochthonous, 41
Azolla, v, vii, 65, 66, 68, 73, 75, 77, 78, 79, 80, 81, 82, 83, 84, 85, 86, 92, 93, 94, 95, 96, 97, 98, 99, 100, 101

B

Bacillus, 52, 55, 56
bacterial microbiota, 37, 42, 58
bacterial species, 37, 40, 44
bactericidal, 56
bacteriocin-like substances, 38, 56, 62
bacteriocins, vii, 38, 39, 55, 56, 60, 61, 62, 63, 64
bacteriostatic, 56
Bacteroides, 40, 43, 44, 47
Bifidobacterium, 40, 44, 46, 47, 49, 57, 58, 62, 63, 64
bulgaricin, 56

C

caecum, 41
CFU/mL, 40, 50
chromosomes, vii, 177, 178, 179, 181, 188, 189, 190, 192, 193
chronic atrophic gastritis (CAG), 38, 43, 44, 46
chronic gastritis, 40, 45, 53, 57
Citrobacter, 45
clarithromycin, 46, 51, 52, 53, 54
clinical trials, vii, 39, 53, 57
Clostridium, 40, 43, 44, 45, 46, 47, 48, 51
coliform bacteria, 40
colon, 37, 40, 41
colonization, 43
cytoplasm, 1, 2, 11, 12, 13, 14, 15, 18, 19, 67, 114

D

danger, 1, 16, 18, 19
dermis of the skin, 1, 3, 16
distribution in onions, 177
duodenum, 40, 43, 45, 47, 64
dysbiosis, 38, 57
dyspepsia, 53

E

electrome, 134, 135, 136, 140, 142, 143, 153, 154, 157, 161, 162, 165, 166, 167, 169, 173, 176
Enterobacteriaceae, 44, 45, 46, 47, 48

Enterococcus, 37, 41, 43, 47, 61, 63
entire biocenosis, 1
eradication rate, 38, 44, 53, 54, 57, 63, 64
Escherichia coli, 43, 47, 56
Eubacterium, 40, 43, 44, 45, 46, 47
eukaryotes, 1, 2, 14, 19, 115, 184
expansion, v, vii, 1, 12, 18, 19, 137, 193
expansion of the bacterium, 1

F

feces, 42
furazolidone, 52, 53
Fusobacterium, 40, 45, 46, 47

G

gastric adenocarcinoma, 39
gastric cancer, 40, 45, 48, 53, 57
gastrointestinal tract (GIT), v, 37, 38, 39, 43, 57, 64
GI tract, vii, 37, 39, 40, 41, 43, 45, 47, 57, 63
growth, v, 38, 57, 66, 68, 70, 73, 75, 76, 77, 78, 80, 83, 92, 93, 94, 96, 97, 99, 101, 103, 104, 105, 110, 111, 112, 113, 118, 120, 122, 123, 124, 125, 127, 128, 129, 131, 150, 163, 165, 169, 172, 174
gut microbiome, 42, 62, 64
gut microbiota, vii, 38, 39, 40, 57, 58, 59, 61

H

Haemophilus, 45, 48
health-promoting properties, 39
Helicobacter pylori (H. Pylori), v, 37, 39, 40, 58, 59, 60, 61, 62, 63, 64
high fat diet (HFD), 38, 43, 47
histomorphometric, vii, 23, 24, 25, 26, 27, 32
host, 34, 38, 39, 53, 137, 142, 152

I

ileum, 40, 41
indigenous intestinal flora, 39
intestinal microflora, 39, 44, 60

J

jejunum, 40, 41

L

Lachnoanaerobaculum, 45, 47
lacticins, 56
Lactobacillus, 37, 38, 39, 40, 41, 43, 44, 45, 46, 47, 48, 49, 53, 58, 59, 60, 61, 62, 63, 64
Lactobacillus GG (LGG), 38, 49, 50, 53, 63
Lactococcus, 37, 41, 47, 57, 60
lansoprazole, 52, 53, 54
leucocin, 56
Leuconostoc, 37, 41, 61
lipid, v, vii, 26, 70, 93, 94, 95, 96, 103, 104, 105, 106, 111, 112, 113, 114, 115, 116, 117, 118, 119, 120, 121, 122, 123, 124, 126, 127, 128, 129, 130, 131, 132
lipid metabolism, v, vii, 103, 104, 105, 111, 113, 116, 117, 119, 130, 132
Listeria monocytogenes, 56
live microorganisms, 39, 53
liver, 1, 3, 16, 17
long-distance signaling, 134, 144, 157
long-range order, 2, 177, 178

M

MALT lymphoma, 39
metagenomes, 42
methyl tertiary butyl ether (MTBE), v, vii, 23, 24, 25, 26, 27, 28, 29, 30, 31, 32, 33, 34, 35
micro biocenosis of human saliva, 1
microalgae, 103, 104, 105, 106, 107, 108, 109, 111, 112, 113, 114, 115, 116, 117,

120, 121, 122, 123, 124, 125, 126, 127, 128, 129, 130, 131, 132
microbial communities, 39, 42, 46, 47, 48, 57
microbial diversity, 41, 43, 57, 61
microbial microbiota, 38
Micrococcus, 44
mitDNA, 177, 178, 184, 186, 187, 189, 191, 192
mucosa associated lymphoid tissue (*MALT*), 38, 39

N

Neisseria, 45, 48
nisin, 56
novel therapeutics, 38, 57
nucleosomes, 21, 177, 178, 180, 181, 183, 184, 187, 188, 189, 190, 192

O

Olsenella, 45, 47
omeprazole, 44, 47, 51, 52, 53, 54, 58

P

Parvimonas, 45, 47
pathogens, 39, 64, 70, 138, 139, 159
patients, 43, 45, 48, 53, 54, 57, 151
pediocin, 56
peptic ulcer, 39, 40, 53, 57
Phyllobacterium, 45, 48
phytoremediation, v, vii, 65, 66, 73, 75, 79, 81, 84, 86, 93, 94, 96, 97, 98, 99, 100, 101, 122
placebo, 49, 50, 51, 52, 53, 54, 62, 63
plant defense, 70, 134, 145, 149, 152, 155, 158, 160, 162
plant electrophysiology, 134, 152, 158, 160, 162, 164, 166, 167, 169, 170, 171
plant-insect interaction, 134, 136, 138, 159, 163
pollutants, 65, 66, 73, 75, 84, 103
potential danger, 1
Prevotella, 43, 44, 45, 46, 47, 48
probiotics, vii, 38, 39, 51, 53, 54, 57, 58, 59, 60, 62, 63
Propionibacterium, 44
Proteobacteria, 43, 44, 45, 46, 47
Proteus, 40
proton pump inhibitor (PPI), 37, 38, 39, 43, 44, 46, 49, 51, 53, 55
protozoa, 40, 104
Pseudomonas, 40, 56

Q

quadruple therapeutic regimen, 51, 55

R

R *radiobacter*, v, vii, 1, 2, 3, 6, 8, 9, 10, 18, 19, 20, 193
recto-sigmoid colon, 41
regimen, 37, 52, 53, 54, 63
Rhodococcus, 45, 48
ribosomal DNA (rDNA), 38, 41, 59
Ruminococcus, 43, 47

S

Saccharomyces, 51, 55
Salmonella Typhimurium, 56
sequential treatment, 54
sex hormones, vii, 23, 24, 26, 30, 32, 34
sex ratio, v, vii, 23, 24, 25, 26, 27, 30, 31, 33, 34, 35
small intestine, 40, 62
solenoids, 177, 180, 183, 185, 187, 189, 192
spent culture supernatant (SCS), 38, 56
Staphylococcus, 40, 43, 47, 56
stomach, 37, 40, 41, 42, 45, 47, 48, 58
Streptococcus, 37, 40, 41, 44, 45, 47, 48

T

Tessarococcus, 45, 47

testis, v, vii, 23, 24, 25, 26, 27, 28, 29, 30, 32
Ti-plasmid, v, 1, 2, 3, 4, 5, 6, 7, 10, 11, 12, 13, 14, 15, 16, 17, 18, 19, 191, 193
tolerance mechanism, 66, 68, 70, 78, 80, 85, 92
triple therapy, 53, 54, 59, 63

V

variation potential (VP) (slow wave potentials (SWP)), 134, 135, 140, 145, 147, 148, 149, 155
Veillonella, 40, 44, 45, 46, 47
Verrucomicrobia, 43, 44, 46, 47
vonoprazan, 51, 55

W

water, 1, 2, 3, 5, 10, 11, 12, 13, 14, 15, 16, 17, 18, 19, 20, 23, 24, 34, 65, 66, 71, 72, 73, 93, 94, 95, 97, 98, 99, 100, 101, 110, 112, 125, 131, 141, 149, 158, 161, 170, 185, 189
Weissella hellenica, 56, 64

X

Xanthomonadaceae, 45, 48

Y

yeasts, 40, 44